Wageningen UR Frontis Series

VOLUME 14

Series editor:

R.J. Bogers
Frontis – Wageningen International Nucleus for Strategic Expertise,
Wageningen University and Research Centre, Wageningen, The Netherlands

Online version at http://www.wur.nl/frontis

THE AGRO-FOOD CHAINS AND NETWORKS FOR DEVELOPMENT

Edited by

RUERD RUBEN

Development Economics Group, Wageningen University,
Wageningen, The Netherlands

MAJA SLINGERLAND

Sustainable Development and System Innovation Group,
Wageningen University, The Netherlands

and

HANS NIJHOFF

International Agricultural Centre,
Wageningen University, The Netherlands

 Springer

A C.I.P. Catalogue record for this book is available from the Library of Congress.

ISBN-10 1-4020-4592-1 (HB)
ISBN-13 978-1-4020-4592-9 (HB)
ISBN-10 1-4020-4600-6 (PB)
ISBN-13 978-1-4020-4600-1 (PB)

Published by Springer,
P.O. Box 17, 3300 AA Dordrecht, The Netherlands.

www.springer.com

Printed on acid-free paper

Printed in the Netherlands.

CONTENTS

Business cases

Summary and conclusions

PREFACE

"From seed to dish, from poor to rich". This was the title of the speech I gave on behalf of the Dutch Minister for Development Cooperation at the opening of the international conference on *Agro-food chains and networks as instruments for development*. This one-liner summarises the relevance of the conference to development cooperation. Why are supply chains important for poverty reduction?

A link needs to be made with the Millennium Development Goals (MDGs). MDG1, 'Eradicate extreme poverty and hunger' and MDG 8, 'Develop a global partnership for development' are directly relevant when discussing chains as instruments for poverty reduction. Achieving the MDGs is not the sole responsibility of the public sector, however. Private companies have a role to play as well and that is why public-private partnership is a recurring theme in contemporary development policy. It means working together towards common goals and sharing the risks, responsibilities, resources, competences and rewards. Management of supply chains is crucial as it requires co-operation between stakeholders.

The conference provided a wealth of examples illustrating that supply-chain cooperation offers potential advantages compared to buying and selling on the open market. The chapter 'Food chains and networks for development: lessons and outlook' contains many recommendations for public and private involvement and also for research derived from these examples.

I would like to draw attention to two broad sets of recommendations related to the donors' role. Firstly, donors can support the public sector in improving conditions in the chain environment. I am thinking of trade-enhancing aspects such as market organisation, competition policy and the operation of public service-agencies. Donors can stimulate coherence in policy making for instance between trade and public health requirements, to help break down non-tariff trade barriers. Secondly, donors need to contribute to strengthening capacity building and training of smallholder farmers. They could also facilitate third party involvement to encourage trust building and control within supply chains as serious options for further support.

A growing number of donors already contribute to enhancing the business climate, for example by improving the regulatory framework, supporting trade facilitation and compliance with public and private certification and standards. The lessons learned demonstrate that the conference provided a good opportunity to reflect on these policies. It is clear that the exchange of information and the dialogue with stakeholders that took place at the conference offered a valuable opportunity to assess existing policies, sharpen visions and approaches, and provided input for policy making.

However, during the conference it was also pointed out that the above mentioned advantages cannot be obtained without major costs and efforts. Moreover, the achievement of an equitable distribution of costs, efforts and benefits is a major issue at stake. The editors of this book of proceedings refer to this as 'equitable integration of developing countries' producers into sustainable (inter)national agro-food chains'. The recommendations address the conditions and critical factors required for inclusive agro-food chain development. Nevertheless, they do not fully address the question of how to harness the full potential of supply chains in terms of pro-poor growth.

For development co-operation, however, answering the "how-question" is fundamental, so this is a challenge for all parties involved on the road to pro-poor value chains. Research, training, stakeholder co-operation and information exchange are needed to contribute to future policy recommendations. Perhaps this will be the theme of a future conference. My compliments to Wageningen University and Research Centre for organising this conference, an important step on this road.

Rob de Vos
Deputy-Director-General for International Cooperation
Ministry of Foreign Affairs
The Netherlands

INTRODUCTION AND ANALYTICAL FRAMEWORK

CHAPTER 1

AGRO-FOOD CHAINS AND NETWORKS FOR DEVELOPMENT

Issues, approaches and strategies

RUERD RUBEN, MAJA SLINGERLAND AND HANS NIJHOFF

Wageningen University and Research Centre (WUR), P.O. Box 9101, 6700 HB Wageningen, The Netherlands

Abstract. Agro-food chains and networks play an increasingly important role in providing access to markets for producers in developing countries. Globalization of trade and integration of supply chains lead to new demands regarding food quality and safety. Analytical approaches for addressing the role of trade for development involve a mixture of disciplines that focus on issues of efficiency, organization and innovation as key dimensions of competitiveness. Smallholder participation in global supply chains is critically determined by three processes: market access, network governance and chain upgrading. Public and voluntary agencies may provide important contributions for reinforcing the supply-chain environment.

Keywords: globalization; international trade; supply-chain integration; network cooperation

INTRODUCTION

Globalization, urbanization and agro-industrialization put increasing demands on the organization of agro-food chains and networks. Food and agribusiness supply chains and networks – once characterized by autonomy and independence of actors – are now swiftly moving toward globally interconnected systems with a large variety of complex relationships. This is also affecting the ways food is produced, processed and delivered at the market (Reardon and Barrett 2000; Van der Laan et al. 1999). Perishable food products can nowadays be shipped from halfway around the world at fairly competitive prices. The market exerts a dual pressure on agro-food chains, forcing towards continuous innovation and agency coordination. Classical price and quality issues are more important than ever, since consumers can choose from an increasing number of products offered by competing chains.

The increasing integration of local and cross-border agro-food chains can be considered both a threat and a challenge for rural development. Poor farmers in developing countries who have limited resources and scarce access to markets and

R. Ruben, M. Slingerland and H. Nijhoff (eds.), Agro-food Chains and Networks for Development, 1-25.

information meet major constraints for the adoption of technological innovations and may therefore be excluded from trade. Economies of scale in processing, transport and distribution also lead to demands for growing volumes of commercial agricultural production and stable delivery capacities of homogeneous quality. Otherwise, smallholder production could offer cost advantages for the delivery of labour-intensive commodities that require strong quality supervision. Involving family farmers into global agro-food chains would also be a suitable device for ensuring a more equitable distribution of the value-added. Bridging the gaps between local economic development and global chain integration asks for the emergence of new institutional and organizational networks that enable producers in developing countries to meet business requirements and trade standards. It also requires a fundamental reorganization of information streams and agency relationships, providing opportunities to smallholders to adjust their supply to consumers' demands and to become a recognizable part of global sourcing regimes.

In this introduction we synthesize main issues at stake in the debate on the role of agro-food chains and networks as instruments for development. First, we summarize the implication of globalization and market liberalization for the organization of local and global food chains. Thereafter, we outline the main principles and approaches that motivate a paradigm shift towards more integrated and interdisciplinary agro-food chain and network analysis. This is followed by a discussion on the institutional aspects of chain and network cooperation. Next we identify the necessary conditions for successful and equitable integration of developing countries' producers into sustainable agro-food chains and networks. We conclude with some implications for policy support to foster entrepreneurship, co-innovation and cooperation between local producers' networks and (inter)national agro-food business companies.

GLOBALIZATION AND INTEGRATION OF AGRO-FOOD CHAINS

Food and agribusiness chains are greatly affected by consumers' concerns regarding food quality and safety and the sustainability of food production and handling methods. Societal concerns regarding GMOs, chemical residues and environmental impact have to be met in a competitive, increasingly global environment. Higher consumer demands regarding the quality, traceability and environmental friendliness of products and processes call for fundamentally new ways of developing, producing and marketing products (Humphrey and Oetero 2000; Omta et al. 2001). This triggers the development of grades and standards and agreements regarding good production and management practices, as well as adequate monitoring systems to guarantee prompt responses and quality compliance. Integrated production, logistics and information and innovation systems become of critical importance for maintaining a competitive market position. In order to achieve international collaboration between farmers, agro-industries and retail companies, strategic and cross-cultural alignment, relational trust and compliance to national and international regulations have become key issues. Mutual learning procedures and feed-back mechanisms are important to guarantee such global alliances.

In recent decades, the world has witnessed an increasing integration of developing-country firms into geographically dispersed supply networks or commodity chains. These chains link together producers, traders and processors from developing countries with retailers and consumers in urban centres and in the developed countries (Gereffi and Korzeniewitz 1994). Firms and companies involved in global food and agribusiness chains and networks are facing fast changes in the business environment, to which they must respond through continuous innovation. New procedures and practices for organizing food supply networks – with direct ties between primary producers, processors and retailers – emerged to cope with food safety and health demands. Optimizing the individual stages in a chain usually results in sub-optimal overall chain performance. For this reason, agro-food companies try to enforce regulations to all actors in the chain that become part of the global market and institutional environment (Jongen 2000; Van der Laan et al. 1999). Firms in developing countries face, however, specific constraints related to limited access to (technical and market) information and reduced borrowing opportunities (Harris-White 1999). Chain integration can then be helpful to improve prospects for sustainable resource management based on more stable access to markets and information that enable additional investment in food quality management (Kuyvenhoven and Bigman 2001).

Recent studies regarding trade and development focus attention on emerging barriers to agricultural exports from developing countries due to stringent sanitary and phytosanitary requirements (Henson and Loader 2001; Otsuki et al. 2001). Liberalization of global trade is increasingly accompanied by technical measures that impose quality standards regarding residues, additives and microbiological contamination. In addition, rapid concentration takes place in the retail sectors for food products – both in developed and less-developed countries – where US- or EU-owned supermarket chains (e.g., Royal Ahold, Carrefour, Tesco, Sainsbury's, WalMart) control an increasing share of food supply to urban consumers. Retailers are also devoting more shelf space to convenient high-quality fresh products (self-service) that are crucial to attract and retain middle-class customers (Fearne and Hughes 1998; Marsden and Wrigley 1996). This poses additional demands on producers and processors to satisfy high and uniform quality standards and frequent delivery requirements (Reardon et al. 1999). International sourcing of perishable products to secure year-around supply (under private label) can be guaranteed through partnerships and long-term contracts. Inclusion of smallholders from developing countries into global supply chains that satisfy these conditions used to be based on procedures for outsourcing and sub-contracting under strict surveillance with frequent audit of local facilities and practices (Dolan et al. 1999). In practice, however, an increasing degree of vertical integration within food and agribusiness networks can be noticed, based on complex contractual arrangements for monitoring product quality and process standards. Consequently, producers can only maintain their market position if credible measures are taken to enhance product quality and safety.

The complex linkages between the before-mentioned processes of market integration and globalization, accompanied by tendencies of growing urbanization and changing consumption patterns, bring about a number of fundamental changes in the organization of agro-food chains and networks. The rapid growth of supermarkets (see Box 1) in both developed and developing countries deeply transforms the institutional landscape of agro-food production and exchange systems. Major challenges as how to guarantee the involvement of smallholder producers in these new and more demanding sourcing networks need to be addressed. Attention should also be given to the institutional requirements that enable smallholders to meet the more stringent food safety and quality regulations. International competition is increasingly taking place around the enforcement of (public and private) regimes of grades and standards. Putting the principles of chain reversal in practice implies that innovative approaches are required that address the necessary conditions for successful and equitable integration of developing countries' producers into sustainable agro-food chains and networks that are capable to satisfy these changing consumer demands.

Box 1. *The rapid rise of supermarkets in developing countries*

Consumers in developing countries purchase an increasing share of their daily food through supermarket chains. Retail sales of fresh products already represent 2-3 times the size of agricultural exports. The supermarket share in food retail is estimated between 40 and 70% in Latin America and Asia and 10-25% in Africa, and increasingly involves middle- and working-class segments of the population in (peri-)urban and even rural regions.

Supermarket procurement regimes for sourcing of fruits, vegetables, dairy and meat strongly influence the organization of the supply chain. The market requires product homogeneity, continuous deliveries, quality upgrading and stable shelf life. Procurement reliance on wholesale markets is rapidly replaced by specialized wholesales, subcontracting with preferred suppliers and consolidated purchase in regional warehouses. Supermarkets thus increasingly control downstream segments of the chain through contracts, private standards and sourcing networks.

Source: Reardon and Timmer (in press)

TRADE AND DEVELOPMENT: TOWARDS A NEW PARADIGM

Early studies on the role of international trade for development have focused on cross-country assessments of the terms of trade and provide recommendations to public agencies regarding appropriate exchange-rate regimes and conducive monetary policies (Krueger et al. 1988). In a similar vein, economic integration has been envisaged from the perspective of creating free-trade zones amongst neighbouring countries. The competitive advantage of most developing nations was considered to be based on their natural resource endowments (i.e., favourable climate conditions for growing tropical crops) and their low relative land and labour costs. Foreign direct investments are mainly channelled towards those developing countries that maintain stable economic performance, provide a reliable legal and fiscal framework, and possess adequate infrastructure facilities.

Porter's (1990) seminal study on the 'Competitive Advantage of Nations' marks a shift in the analysis of trade and economic development, focusing attention at the competition and cooperation among enterprises instead of countries. In his view, competition increasingly takes place between firms and amongst supply chains that try to improve their position through systems upgrading and superior management regimes. This has far-reaching implications for development studies, since more attention should be given to the interfaces and linkages between farms and firms. Private-sector-oriented marketing studies and agribusiness analyses thus conquered a new space in the development arena (Cook and Chaddad 2000).

Research on marketing of smallholder crops in developing countries has traditionally been strongly supply-driven, focusing attention on 'finding market outlets' (Scott 1995) while paying scarce attention to consumers' demands. Most early studies on international trade refer to course grains and staples and focus on the efficiency of traders and collectors networks. Chain cooperation was usually limited to the delivery contracts, considering external relations within the framework of interlocked transactions and sub-contracting arrangements (Glover 1990; Key and Runsten 1999). Integrated analyses of international commodity chains have focused on long chains with considerable value-added in transport and processing (e.g., coffee, cotton, sugar, bananas; see Vellema and Boselie 2003; Dorward et al. 1998; Van der Laan et al. 1999). Some studies on fair trade and ecologically produced commodities are confined to particular market niches (e.g., FLO and IFOAM certification).

On the other end, agribusiness analyses usually devote limited attention to the existing trade-offs between consumers' food demands and producers' welfare. Spot-market exchange or loose delivery contracts are not able to bridge this gap. Given the increasing globalization of transactions in fresh products, new market institutions emerge that better respond to the dynamics of agro-food systems. Promising analytical frameworks making use of agency theory and contract choice simulation have recently become available that permit to identify potential win-win scenarios. Improved integration of global commodity chains is increasingly considered a suitable strategy for enhancing food quality and sustainable resource management practices at different scale levels. Under conditions of market liberalization, contractual relations may offer alternatives for simultaneously enhancing food safety standards and reducing risks (Van Tilburg and Moll 2000).

New concepts

Supply-chain analyses make use of a range of concepts to identify critical aspects of market structure and performance. *Supply chains* are understood as transformation processes from inputs through primary production, processing and marketing to the final consumer (Porter 1990). They involve three key dimensions: (a) organizational systems for the coordination amongst agents; (b) knowledge systems for combining information, skills and technologies; and (c) economic mechanisms for product and technology selection and for providing market access. Supply chain performance can be assessed with efficiency parameters, searching for specialization according to comparative advantage and towards integration for reducing transaction costs.

Additional performance indicators (Beers 2001) are in the domain of consumer value (e.g., perceived quality) and impact on society (e.g., side effects on environment and health).

The French '*filière*' (or sub-sector) approach – defined as a system of agents for producing and distributing goods and services – provides insight into the sequential nature of interconnected activities through the spatial mapping of commodity flows. Main attention is given to the empirical assessment of input–output relations, prices and value-added distribution along commodity chains (Raikes et al. 2000). Commodity systems are mostly analysed from a rather linear technical and managerial perspective and *filière* analyses have been widely used to justify commodity price stabilization regimes.

Value chains focus attention on the distribution of value-added throughout the supply chain amongst different agents (Gereffi and Korzeniewitz 1994; Gereffi et al. 2002). This analysis devotes special attention to the cost structure of production, processing, transport and retail, the opportunities for reaching economies of scale and scope, and the available surplus that accrues to each of the chain partners. This value distribution is subject to bargaining amongst the chain partners and will be modified when increasing interdependencies give rise to changing perceptions of risk and efforts. Analyses of global commodity chains devote particular attention to the governance dimension of trade networks, the existence of entry barriers and the economic and spatial division of labour.

Relations between partners involved in the supply chain can be analysed with different concepts. Spatial cooperation has been addressed through *clusters* that consist of a geographical concentration of interconnected activities with strong vertical linkages in order to reinforce competitiveness (Porter 1998). The advantages of clusters involve economies of scale and scope, providing opportunities for flexible specialization to reduce technological discontinuities, and agglomeration effects that permit lower transaction costs. Clusters thus create external economies (i.e., labour and input exchange; joint learning; reduced transport costs) and reinforce collective efficiency through collective action in areas of mutual interest.

In a similar vein, *networks* are envisaged as horizontally structured relationships between agents that enable a reduction of transaction costs for coordination and information exchange. Agency coordination permits the creation of scale economies for input purchase and marketing, complementarities in the division of tasks, and network externalities (Hayami and Otsuka 1993). Taking advantages of the existing diversity in resources and capacities, networks based on pooled interdependence can thus reinforce the bargaining position of agents within the chain.

Recently, Lazzarini et al. (2001) launched the concept of *netchains* at the interface of vertical supply chains and horizontal networks. Netchains can be conceptualized as a multi-layer hierarchy between suppliers, processors and retailers where horizontal coordination between reciprocal agents is embedded in a framework of vertical deliveries (see Box 2). Horizontal cooperation (e.g., in farmers cooperatives) may be better able to cope with the stringent quality criteria and changing quantity demands emerging from chain partners.

Box 2. Example of a netchain structure

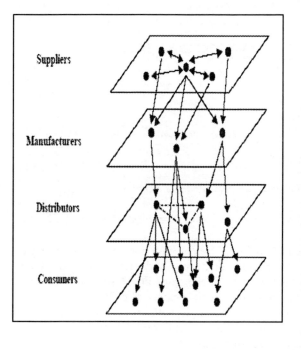

Suppliers

Manufacturers

Distributors

Consumers

Netchains provide linkages between horizontal networks of suppliers and vertical supply chains. They involve different types of (nested) interdependencies amongst agents, like: (a) reciprocal cooperation based on mutual exchange between suppliers; (b) sequential delivery systems based on planning along the supply chain; and (c) pooled interdependencies at business level to guarantee standardization and harmonization of processes.

Source: Lazzarini et al. (2001)

Finally, *contracts* play a critical role in the relationships between chain and networks partners. They define the rules and obligations for establishing cooperation, both between network partners and chain agents. When repeated transactions take place, contracts represent a cost-reducing device. For deliveries that involve high-quality demands, self-enforcing contracts that involve trust and loyalty are preferred to reduce monitoring costs. Different options for integrating (horizontal) networks and (vertical) chain contracts are available for guaranteeing risk-sharing and ensuring trust relationships. Given the high risks and the difficulties of monitoring numerous heterogeneous agents, entire-channel process control is increasingly preferred (Van der Laan 1993; Janssen and Van Tilburg 1997).

INTERDISCIPLINARY PERSPECTIVES ON AGRO-FOOD CHAINS

Supply-chain analysis is becoming an interdisciplinary activity. Production and distribution processes involve a mixture of socioeconomic, technological, legal and environmental criteria that are highly complementary in explaining overall agro-food chain performance (see Figure 1).

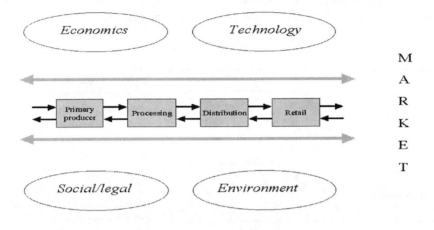

Figure 1. Analytical perspectives on food chains

The performance of the entire food chain *'from farm to fork'* is shaped by four different dimensions (Trienekens 1999):

- Economic dimension, related to chain efficiency (in a cost–benefit perspective) and consumer orientation. To increase efficiency and profitability, individual companies may establish alliances with other parties in the production column resulting in supply chains and networks. Such 'netchains' offer better prospects that production and distribution systems comply with consumer values, enable the establishment of integrated quality and safety control systems, and might enhance the external competitiveness of businesses.

- Environmental dimension, referring to the way production, trade and distribution of food is embedded in its (ecological) environment. Important performance issues are related to the use of energy and to energy emissions in production and distribution of food products, the recycling of waste and packaging materials throughout supply chains, and the prospects for sustainable food production systems (including attention for issues like biodiversity and landscape architecture).

- Technological dimension, related to the application of (product and process) technology, logistical systems, and information and communication technologies that improve quality performance and enhance innovation in food products. Important issues at stake refer to systems for guiding and controlling processes and flows of goods throughout the supply chain (e.g. HACCP, tracking and tracing) and the development of new products supported by (private) standards.

- Legal and social dimension, i.e. the norms and values related to societal constraints to production, distribution and trade of food, concerning criteria of human well-being, animal welfare and sustainable entrepreneurship. Important issues at stake refer to legislation and agreed business practices (in platforms and conventions) regarding food products, compliance with corporate social responsibility (People–Planet–Profit), and the (inter)national legal and regulatory framework.

Central aspects influencing the performance of food supply chains are usually found at the interface of private and public action. Consumer expectations and demands regarding food quality and safety can be addressed through technological optimization (e.g., improved integration of production and distribution systems to reduce delivery times and improve shelf life), with specific management practices (brands, informational labelling, etc.) accompanied by suitable monitoring and control systems (traceability), and/or by imposing legal standards. Similarly, the sustainability of food chains can be enhanced through technical interventions (improved seeds, biodiversity management, waste disposal), with private economic measures (environmental labelling, differentiating food products complying with particular health and safety standards), within the framework of (inter)national legal standards and socio-cultural customs.

New approaches for agro-food chain studies

In recent years, important progress has been made in the development of new approaches for analysing the structure and dynamics of agro-food chains and networks (Lazzarini et al. 2001; Omta et al. 2001). Scientific approaches that contributed to the innovation of supply-chain and network analysis can be grouped into three main traditions:

- Supply-chain management (SCM) as a customer-oriented approach that aims at the integration of business planning for balancing supply and demand across the entire supply chain (Bowersox and Closs 1996; Cooper et al. 1997). Advanced information and communication technology systems are increasingly becoming the backbone of integrated supply chains (Lancioni et al. 2000; Porter 2001). Supply-chain management research is supported by mathematical modelling and simulation tools (Van der Vorst 2000; Trienekens and Hvolby 2001).Within SCM total quality management (TQM) and assurance systems such as good agricultural practices (GAP), good manufacturing practices (GMP), ISO and hazard analysis and critical control point (HACCP) gain importance (Luning et al. 2002). GAP, GMP and HACCP focus mainly at technology and ISO at management. TQM strives for continuous improvement in all functions in an organization based on a quality concept that is based on management commitment and employee empowerment and utilized from acquisition to service after sales (Kaynak 2003).

- Network and contract choice (NCC), where the necessity for organizations to exchange resources is a key factor for inter-organizational relationships (Håkånsson and Snehota 1995). In network theory, forms of collaboration are not only based on economic motivations, but power and trust are equally important (Uzzi 1997). Social-capital theory has become an important new branch within the network approach. Network relations may enhance the 'social capital' of a company through improved access to information, technical know-how and financial support (Coleman 1990; Burt 1997; 2002). Empirical approaches for analysing interfaces between agents within a network draw on contract choice theory (Hayami and Otsuka 1993). Making use of classic models for sharecropping, attention is focused on interlinked exchange transactions at input and commodity markets that respond to certain product or process standards and satisfy delivery conditions, while reducing monitoring costs and risk. Modern applications of contract choice also embrace chain quality management aspects (Weaver and Kim 2001) and loyalty issues (Saenz and Ruben 2004).

- The new institutional theory of transaction-cost economics (TCE) and agency theory provides the rationale for make-or-buy decisions (Rindfleisch and Heide 1997; Williamson 1987; 1999). These approaches are concerned with governance regimes for organizational cooperation, integrating views from business economics and organizational theory. Agency theory is directed at the ubiquitous agency relationship, in which one party – the principal – delegates work to another – the agent – who performs that work (Eisenhardt 1989).

Recent approaches devote major attention to the interfaces between technical and institutional strategies for overcoming the classical trade-off between (a) investments in improved product standards and process management practices; and (b) the derived value-added and income-generation effects at different stages of the commodity chain. Reduction of transaction costs and risks can be reached through improvement of the effectiveness in contract compliance between different agents involved in the chain (Sheldon 1996). Monitoring food safety increasingly depends on vertical coordination and contracting mechanisms that involve all relevant partners, with complementary roles for public and voluntary agencies (Antle 1996). These approaches make efforts for linking consumers' demands regarding food safety attributes and sensory preferences with producers' and processors' practices within the framework of global network governance and international chain integration.

CHAIN AND NETWORK COOPERATION

In an increasingly globalizing world, the organization of the agro-food sector is subject to rapid change. The institutional structure and governance regimes within global value chains are shaped by a series of structural changes that substantially modify the production and exchange relationships. We highlight the most important trends in supply-chain governance that are relevant for developing countries.

Buyer-driven chains

Chain cooperation has traditionally been based on producer firms that started to manufacture commodities in overseas factories. Foreign direct investments were focused on primary production and processing, while major concentration took place in upstream segments. In recent years, global buyers and retailers have begun to play a key role in the integration of production and distribution networks (Reardon and Timmer in press). Market access is highly dependent on participating in such global supply networks. Traditional commodity chains are also becoming more differentiated (Fitter and Kaplinsky 2001). Rapid adjustment to changes in consumers' demands has become a key element of competitiveness.

Contractual governance

Spot-market relationships that were guided by prices are increasingly replaced by governance regimes characterized by hierarchy and managerial control (vertical integration). In addition, network governance through contractual relationships between autonomous firms is guided by complex sets of delivery arrangements, where price and non-price elements are equally important. Gereffi et al. (2002) distinguish between relational networks mediated by trust and reputation, modular networks using standards and information as coordinating mechanisms, and captive networks organized around monitoring and control. Innovative networks are characterized by multi-polar governance structures where potential drivers are located in different nodes of the chain. Reputation, trust and loyalty have become critical to guarantee effective governance.

Innovation through alliances

The locus and character of innovation processes are subject to important change. Instead of simple technology transfer, research and development activities now involve both technological and managerial dimensions. The processes of product development and upgrading are increasingly structured as co-innovation activities that take place in alliances between chain partners. This involves close linkages between 'hardware' (production, processing and logistics) with 'software' (organization, management) through expertise development based on the exchange of experiences with chain and network partners. International competition asks for a continuous learning through reorganization of production processes and network upgrading with strong interactions between design, production and marketing operations.

Continuity and flexibility

Supply-chain organization has become strongly oriented towards criteria of continuous delivery and flexible sourcing. Continuity is of vital importance to guarantee shelf space, while repeated transactions in the supply chain enable the establishment of reputation and trust. Logistical systems are optimized in order to reduce costly stock-keeping operations while avoiding out-of-stock. Forecasting of demand and flexibility in sourcing regimes – including options for global sourcing – enable retailers to guarantee year-round supply of perishable products at more or less stable prices.

Information and communication

Profits and trade margins increasingly depend on information flows regarding customer demands for design, packaging, distribution and servicing of products. Keesing and Lall (1992) point to critical information requirements that enable firms to improve competitiveness and market responsiveness. In addition to product information, retailers have to respond to consumers' concerns regarding food safety, labour standards and environmental effects. Control and compliance with these issues are ensured through entire chain monitoring, based on tracking and tracing information systems (see Box 3).

Box 3. *Eurepgap*

The branch organization of European retailers (Eurep) established a code for 'good agricultural practices' (GAP). Developing-countries producers have to fulfil a list of technical, handling and managerial practices to guarantee quality, consistency, hygiene and safety. Through regular inspections and the use of bar codes, a system of tight coordination is installed that enables entire supply-channel information and control. Local producers have to make substantial investments for complying with these rules, but only a limited number of producers acquire the preferred supplier status.

Grades and standards

The role of grades and standards (G&S) has shifted from a technical instrument to reduce transaction costs in homogeneous commodity markets towards a strategic instrument of competition in more differentiated product markets (Reardon et al. 1999) In addition, G&S have shifted from performance criteria related to product characteristics to process standards involving all chain operations, to assure consumers of the quality, safety and environmental and/or social characteristics of production and handling practices in distant locations. Finally, private labels, certificates and standards created and enforced by large international retail and agro-food companies are far-ahead public rules, enabling firms to create specific market segments and capture additional rents (Farina and Reardon 2000).

ROLE OF CHAINS AND NETWORKS FOR DEVELOPMENT

International partnerships for sustainable food production and poverty alleviation increasingly pay attention to the organization and performance of agro-food chains and networks. Improving market access and competitiveness of smallholders in developing countries requires concerted efforts for linking different stakeholders (producers, traders, processors and retailers) in order to reduce transaction costs and to reinforce learning capacities. Meeting the market requirements of scale, reliable supply, loyalty and quality is critically important for reaching competitiveness.

Market forces urge supply-chain partners towards closer cooperation. Especially for local producers in developing countries who wish to participate in regional or global markets, supply-chain collaboration is of key importance for guaranteeing:

- <u>access</u> to new and profitable market outlets, based on supply-chain management for innovative product–market combinations;
- <u>network governance</u> for enabling timely responses to demands for capacity development and knowledge dissemination; and
- <u>chain upgrading</u> through partnerships that increase the size and distribution of value-added through improved production systems, information regimes or logistics.

The aspects of market access, governance and upgrading are commonly recognized as the three key dimensions to create opportunities for linking developing countries' producers to dynamic (local and international) markets.

Market access

Falling trade barriers do not automatically lead to better market access for developing-countries firms, especially when supply chains are governed by a limited number of buyers. African smallholders are easily de-listed from vertically-structured horticulture supply networks oriented at European supermarket outlets (Dolan and Humphrey 2000), but can equally become marginalized in local delivery regimes (Boselie 2002).

European imports of fresh food and specialty vegetables (i.e. sugar snaps, baby corn, asparagus, etc.) have increased by 140% in value terms between 1989 and 1997, and sub-Saharan countries were able to capture a consistent 30% of the market share (Humphrey and Oetero 2000). What started as an off-season trade in temperate vegetables and specialist imports for the ethnic market has become a major all-season business.

Making supply chains work for development implies that local producers should not only be cost-competitive, but also able to comply with quality requirements, guarantee constant and reliable supply, and strictly maintain safety and health regulations. While family-operated smallholder farms usually exhibit advantages for producing labour-intensive products (Key and Runsten 1999; Dries and Swinnen 2004), the increasing capital demands for establishing processing facilities, cool chains and logistics systems tend to favour sourcing from larger firms, where inspection and monitoring benefit from economies of scale and scope. Supporting smallholder participation in supply chains not only requires initial market access, but particular attention should be given to consistency, e.g., the capacity to maintain constant deliveries and reliable and uniform appearance, taste and quality over time (Dolan and Humphrey 2000).

Different dimensions of market access deserve special attention. Since economies of scale in food processing and trade are usually larger than in primary production, upstream 'pooling' of farmers through different forms of cooperative associations and networks has become of utmost importance.

Whereas competitiveness may be initially derived from resource and location advantages, access to market information is becoming a main dimension for maintaining competitive advantage. Entry into international markets also requires that due attention is given to delivery and packaging standards that constitute key elements for maintaining any comparative advantage.

For acquiring market access various strategies can be pursued that rely on distinct marketing channels. Van der Laan (1993) distinguishes between (a) entire-channel crops (mainly perishables), where direct contacts and strict coordination between producers and importers are critical for quality assurance; and (b) half-channel crops (standardized products) that split the chain into different segments between producers and exporters. The latter option may initially provide somewhat better opportunities for local smallholders. In addition, most producers rely on multiple market outlets for different quality categories of their production. Once a strong position is gained at the local market, production could be gradually scaled-up towards more demanding (and rewarding) regional or international outlets. Reardon and Timmer (in press) consistently argue that there is still considerable scope for enhancing the competitive position of smallholders in domestic and regional supply chains. Important margins for improving value-added can also be found in strategies for optimizing logistics and information systems (see Box 4).

Box 4. The Dabbawallas network in Mumbai

> Over 200,000 people working across a 70-km stretch around Mumbai city (India) receive every day their lunchbox (*dabba*) through a carry and delivery system operated by Mumbai Carriers Association, a relatively flat organization run and managed by a group of largely illiterate rural working-class people using nothing more than three or four symbols crudely painted on the boxes to guarantee timely delivery. The boxes are home-made and carried by '*wallas*' to hub metro stations where they are reassembled for further transmission via local trains. At the destination, the process of further distribution is spawned. The *dabbawalla* system is based on face-to-face communication where each box changes hands at least four times, but intuition and teamwork guarantee that it operates at very low costs (Rps 100/month) and a surprisingly low error rate (less than 0.5%) for a system of its size.
>
> Source: Kumar et al. (2001)

Network Governance

Given the tendencies of urbanization and globalization, supply chains for agricultural and food products are increasingly challenged by consumers' demands regarding quality and safety. Delivery conditions and procurement regimes also require constant and reliable supply and tend to favour the development of selective preferred-supplier relations. Local smallholders can better compete if embedded in institutional partnerships which enable network coordination and strengthen entrepreneurship in order to pursue a gradual improvement of the terms of trade.

New communication regimes enable business processes to be compressed in time but extended across space. Competition is not only based on production technologies, but far more on new forms of supply-chain organization. The effectiveness of governance networks is strongly related to the establishment of long-term, stable and durable relations between supply-chain partners and a common understanding of shared values. Producers operating at more customized market segments (i.e. certified fair-trade and organic products) also need to organize credible supervision by a third party to ensure that specific production practices are maintained.

Innovation and adaptation are key capacities that need to be developed within suppliers networks. Different types of supply chain operate under governance regimes that provide specific types of incentives for innovation. Within vertically structured delivery chains, the lead firm is fully engaged in the entire range of production activities and exercises strict control over upstream operations (Sturgeon 2001). When firms start relying on subcontracting and outsourcing, most design and product development activities are still maintained by the buyers, and producers frequently need to adapt to changing market demands. Under preferred-supplier regimes, tracking and tracing systems are put in place to guarantee full process control. With increasing technological capabilities in the producing countries – particularly in SE Asia – some local companies acquire greater independence and may eventually become competitors in the market.

Networks strongly rely on agency coordination and tend to be structured in such a way that behavioural and investment risks are controlled. Contracts are increasingly used as instruments to improve product quality and to enforce permanent supply (as well as to define liability in case of substandard deliveries), but trust building is required to guarantee real loyalty and to reduce opportunistic behaviour. Resource-providing delivery contracts proved to be particularly effective in settings where land rents are high and production operations rather labour-intensive, linking smallholding operations with remote markets (Key and Runsten 1999). Co-investment schedules, where private firms – together with banks, state agencies and knowledge institutions – are jointly engaged in supply-chain development, can provide useful leverage for spreading risk and improving the spread of innovations. Finally, companies also started to appreciate transparency and accountability within supply chains as an intrinsic element of their strategies towards corporate social responsibility (CSR and triple P).

Chain upgrading

Agro-food chains nowadays involve considerable processing activities that generate most value-added. Specialized knowledge regarding appropriate inputs, handling practices and logistics is considered of key importance for quality upgrading. Other strategies related to product development and eventually labelling and certification may offer prospects for improving the size and distribution of value-added. In addition, improved access to specific market segments can improve the bargaining opportunities for local stakeholders.

Gereffi et al. (2002) distinguish between four different strategies of upgrading for improving the competitive position of firms: (a) product upgrading; (b) process upgrading; (c) intra-chain upgrading; and (d) inter-chain upgrading. While the first two strategies focus on the development of new products or production systems, the latter two strategies aim acquiring particular competences that enable to start new activities in other market segments or sub-sectors. Successful upgrading proves to be highly dependent on innovative capacities and local institutional support.

Supply-chain management is increasingly considered an important tool for value-added creation. Upgrading strategies can either focus on *diversification* into specific product attributes customized towards particular consumer outlets where premium rates are paid (see Box 5), or be based on market *segmentation* by the labelling of particular products through location-specific branding, packaging or marketing standards. Coffee is a well-known case, where both speciality coffees (gourmet, organic, fair-trade) and branding (Café de Colombia) account for increasing market shares (Fitter and Kaplinsky 2001; Humphrey and Oetero 2000). Upgrading in the fruit and vegetables sector is strongly based on product diversification, but more recently added value is increased through local processing activities (e.g. Ahold fruit salads prepared in Ghana; pre-packed ready-to-eat beans from Kenya for UK supermarkets). A further strategy for increasing value-added emerges when developing-countries producers become shareholders of marketing companies in the North (like in the European fruit company Agrofair), while some UK importers have taken equity stakes in East-African export companies.

Box 5. An indicator system for sustainability in coffee chains

In the coffee sector, important progress has been made for establishing an integrated sector-wide indicator system to assess advances in economic, social and ecological sustainability (focusing on the classic People-Planet-Profit dimensions) and to communicate these achievements to consumers. The broad sustainability concept is translated into audible and measurable indicators, and specific tools and guidelines are developed to enhance the performance of stakeholders in the coffee chain. Assistance is provided to enhance capacities amongst the industry, producers' associations and state agencies for joint implementation of a common code for sustainability in coffee.

Source: Vellema and Boselie (2003)

CRITICAL ISSUES FOR CHAIN AND NETWORK COOPERATION

The effective participation and equitable integration of producers from developing countries in regional and international agro-food supply chains and networks is subject to a wide number of individual competences and institutional constraints. Inclusion or exclusion from production and delivery networks is decisive for way the gains from globalization are spread. We therefore discuss three critical factors that enable developing-countries producers' engagement in integrated agro-food supply chains.

Building experience and trust

Governance in supply chains is exercised through a complex mixture of performance standards combined with behavioural incentives for enforcing compliance. With rising monitoring and auditing costs, building trust and loyalty becomes increasingly important. Since effective coordination within international supply chains turns out to be a cornerstone for maintaining competitiveness, relationships between producers and importers are likely to evolve towards closer interdependence. This is particularly the case when large fixed investments for processing and logistics create asset specificity that can only be contested with long-term delivery contracts (Hueth et al. 1999; Ruben et al. 2004).

Dovetailing learning and innovation

The competitive advantage for agro-food supply chains originating in developing countries is increasingly based on management coordination and adaptive capabilities for responding to changing market demands. Similarly, entrepreneurship is developed through a dynamic process of learning and innovation. Management of innovations within chains and networks requires an interactive process at the interface of customers and suppliers, sometimes also involving knowledge institutions and even competitors (Omta 2004; Håkånsson 1982). Challenging examples of such international co-innovation processes are found in the optimization of logistics and warehouse operations for fruits and vegetables in South Africa and Central America (see case studies included in this volume), and the upgrading of dairy delivery systems in Latin America (Dirven 1999; Farina 2002).

Sharing benefits and rents

The creation of added value is increasingly taking place in the intangible parts of the supply chain, where design skills and brand names are controlled (Kaplinsky 2000). The advantages from integrated supply chains are mainly derived from 'systemic' efficiency where the profits of coordinated action are higher than the returns that can be reached by individual agents. New product–market combinations or improved management procedures generate dynamic rents that are likely to accrue to the most innovative parts of the chain. Distribution of value-added is therefore contingent on the possibilities for engagement in chain upgrading. The development of the *Senseo* coffee-pad technology is a typical example of such technological cooperation between electronics and food industries, shifting the locus of value-added creation to downstream segments of the agro-food system.

STRATEGIES AND POLICIES

Supply-chain management and network governance essentially belong to the private-sector domain. There are, however, several valid reasons for engagement of the public sector and voluntary organizations in improving the chain environment. We outline five main directions of strategic support and potential leverage towards sustainable and equitable chain and network integration, focusing on the complementary roles of public and private agencies.

Reinforcing the business climate

Macroeconomic stability of the exchange rate, control on inflationary pressure and a liberal trade regime are critical investment conditions. In addition, legal protection of (foreign) direct investments and political stability (including corruption control and accountability of tax and trade agencies) represent key elements for establishing integrated agro-food supply chains in developing countries. Public investments for infrastructure provision and social services (education, health care) are equally important to provide an enabling environment for business development.

Given the resource-based character of most agro-food industries, national market integration is key to further growth. Reardon and Timmer (in press) argue that at least 85 percent of food is consumed domestically, and this is particularly true for fresh and perishable products. In recent years, important progress has been made towards spatially and temporally integrated staple markets in most developing countries (Barrett 2001; Badiane and Shively 1998). Price variability of non-staple food products and processed foods is, however, still very large and subject to frequent shocks.

Developing countries are also becoming increasingly involved in international trade of processed foods. Between 1980 and 2002, the value of agro-food exports roughly doubled from $200 to $400 billion (FAOSTAT 2004), but the share of bulk grains dropped from 45 to 30% and major growth was realized in perishables (fruit, vegetables, flowers, fish and meat) and particularly in processed foods (juice, beverages, snacks, etc.) that increased from 18 to 34% (Regmi and Gehlhar 2003 cited by Reardon and Timmer in press). Although foreign direct investments were important to mark this shift, some of these activities were originally oriented towards domestic consumers and gradually 'upgraded' towards regional or international market outlets. Moreover, while in some cases domestic supply was seriously affected (most notably increasing protein deficits due to fisheries exports from Lake Victoria; see Henson et al. 2000), for most other industries domestic outlets still represent an important subsidiary marketing channel.

Establishing the legal framework

Agro-food companies operate within an environment where production practices directly influence consumers' welfare. This implies that there is a legitimate role for public (sometimes semi-autonomous) agencies to exercise control on the maintenance of food safety rules and regulations. In addition to international standards (FAO *Codex Alimentarius,* SPS agreement), also national grades and standards are in place that sometimes compete with private rules for Good Agricultural Practices (GAP). Public regulation may involve normative codes regarding health and safety, but also includes compliance with labour and environmental standards. The latter are strongly advocated by (inter)nationally operating non-governmental organizations, like Greenpeace, IUCN, Oxfam and others. Particular initiatives for labelling fair and ethic trade intend to make food trade more transparent and try to mobilize consumers for these issues. In a similar vein, the private agro-food industry sector has organized a Sustainable Agriculture Initiative (SAI) as an effort towards shaping its corporate social responsibility.

Another aspect of the legal framework refers to ownership rights and the supply-chain governance structure. Apart from the required securities for realizing fixed investments, an important part of supply-chain control nowadays rests in so-called intangible competencies (RandD, design, branding, etc.), which are characterized by high entry barriers and command highest returns (Kaplinsky 2000). As long as the operations of developing-countries firms remain limited to production activities, they are likely to exercise limited governance power and will receive a minor share of the value-added. Joint ventures and strategic alliances between local and international firms may enable producers to acquire business practice and learn best practices. Other options for reinforcing collective action assign a role to business associations in providing market information and monitoring food standards. Finally, sector-wide organization of producers (such as the Fresh Produce Exporters Association of Kenya; FPEAK) may offer prospects for creating countervailing power.

Safeguarding consumers' interest

Governments play an important role in guaranteeing the availability and safety of agro-food products to local consumers. It is therefore in the interest of local consumers that regular inspections take place, and that an acceptable degree of local competition is maintained to guarantee that retail prices are established under competitive conditions. Given the increasing size of domestic markets, the rapid rise of supermarkets in developing countries takes place under intense competition and (poor and middle-class) consumers appear as the main winners (Reardon and Timmer in press), but in the future further concentration in retail and agribusiness may lead to the progressive elimination of small shops and shrinking of wet markets. There is thus certainly room for competition policies that facilitate market entry for (local) producers.

Reducing transaction costs

Guaranteeing the participation of smallholders in agro-food supply chains requires reduction of transaction costs. Market entry is very much dependent on both internal and external economies of scale and scope. Therefore, public provision of road infrastructure and public support for education and training remain critical for overcoming start-up problems. Transport costs and qualification of the labour force are thus becoming key dimensions of the comparative advantage.

Internal economies of scale can be reinforced through decisive efforts towards the establishment of farmers' associations or cooperatives. Notwithstanding the general negative experiences with cooperative production (see Ruben and Lerman 2005; Berdegué Sacristán 2001), farmers demonstrate wide interest in joining efforts for improving market access. Higher food-quality and safety standards can also be better met if farmers make joint investments and are willing to exercise mutual control on free-riding. The latter may provide important cost advantages to small producers who are able to reduce monitoring costs. In addition, cooperatives could exercise bargaining power vis-à-vis traders and retailers and thus gradually improve their share in value-added.

Managing risk

Further engagement of smallholders in (inter)national agro-food supply chains is seriously constrained by risk motives. Fafchamps (2004) provides an extensive overview of the discouraging effects of price uncertainties, risks of product denial and contract breach, and the implications of delayed payments in sub-Saharan markets. Similar evidence on market and price risks in Latin America is presented by Barham et al. (1992).

There is a decisive role to play for public agencies in guaranteeing the legal framework and defining transparent rules for conflict settlement. Farmers can only make the required investments to improve delivery frequency and quality when they can be relatively certain regarding available market outlets. Key and Runstein (1999) indicate that contract farming provides best outcomes under conditions where public surveillance is guaranteed. In addition, some West-African governments have organized market intelligence services to guarantee open access (through radio emissions) to price information. More promising experiences are reached with private-based pre-paid mobile-phone lease facilities in Bangladesh that enable farmers to contact relatives in other places in order to obtain price information (Courtright 2004).

Provision of credit and insurance represents a second major strategy for risk management. Experiments are underway that provide weather insurance to farmers in rain-fed regions upon payment of a fixed hectare fee, thus preventing distress land sales in cases of unexpected harvest losses (Bie Lilleør et al. 2005). More important for supply-chain integration are insurance provisions that are part of the delivery contract (Bogetoft and Olesen 2004). In order to avoid disputes between producers and traders, rules for quality inspection and timely payments need to be sufficiently clear and enforceable. Preferred supplier arrangements may include provisions for cost-sharing and repeated transactions that provide farmers with the required security for making fixed investments.

OUTLOOK

The different contributions included in this volume provide a comprehensive overview of the current state of the art in the field of agro-food chains and networks and their potential contributions to the development process. The book is divided into three parts: (1) a number of analytical papers that address the roles of public, private and voluntary agents in shaping partnerships and alliances that may support market access and permanent supply-chain linkages for smallholders; (2) a series of seven business cases that provide illustrations of particular strategies for supply-chain integration and that identify the critical factors responsible for successful alliances; and (3) three concluding articles that discuss policy implications and provide some strategic guidelines for further action towards the promotion of sustainable and durable network cooperation throughout (inter)national agro-food supply chains.

The articles included in this volume bring together different viewpoints from the public agencies (Roberto Rodriguez, the Brazilian Minister of Agriculture), the business sector (Alfons Schmid from Royal Ahold; Jeroen Bordewijk from Unilever and Johan van Deventer from Freshmark South Africa) and local farmers organizations (Leonard Kariuki from the Kenyan National Federation of Agricultural Producers and Gonzalo la Cruz from the Peruvian Fair Trade Banana Organization). In addition, attention is given to the interfaces between public and private grades and standards (Tom Reardon from Michigan State University) and the role of local agro-food chains and street markets (Olusola Oyewole and Biola Phillip from Nigeria).

The seven selected business cases highlight different dimensions of the organizational structure and management regimes of integrated supply chains originating in developing countries. Cases are presented concerning improved sourcing regimes for supermarket supply of fresh fruits and vegetables in Thailand (Jan Buurma and Joompol Saranark), the design of supply-chain information systems and logistics for fruit exports from South Africa to The Netherlands (Anneke Polderdijk and colleagues), the upgrading of beef supply chains in Brazil (Marcos Neves and Roberto Scare), the quality and management constraints in the Nile-perch supply chain from Lake Victoria (Ronald Schuurhuizen, Aad van Tilburg and Emma Kambewa), the prospects for certification in the organic cocoa chain from Costa Rica (Maja Slingerland and Enrique Diaz Gonzales), the integration of novel supply chains for *Allanblackia* oil in Ghana (Lawrence Attipoe, Annette van Andel and Samuel Kofi Nyame) and the development of supply chains for medical plants in India (Petra van de Kop, Ghayur Alam and Bart de Steenhuijsen Piters).

Finally, the volume provides three concluding chapters that address the challenges for researchers and policymakers. Louise Fresco of the United Nations Food and Agricultural Organization (FAO) outlines what can be done to enhance sustainable agro-food chains through more comprehensive and inclusive standards.

Kees van der Meer of the World Bank provides a detailed overview of the factors that lead to inclusion or exclusion of smallholders from coordinated agro-food supply chains. The editors conclude with a summary of the critical economic, institutional and policy issues that need to be considered in order to guarantee support for smallholder market access, capacity development and functional upgrading that can contribute to dynamic and responsive agro-food chains.

REFERENCES

Antle, J.M., 1996. Efficient food safety regulation in the food manufacturing sector. *American Journal of Agricultural Economics,* 78 (5), 1242-1247.

Badiane, O. and Shively, G.E., 1998. Spatial integration, transport costs, and the response of local prices to policy changes in Ghana. *Journal of Development Economics,* 56 (2), 411-431.

Barham, B., Clark, M., Katz, E., et al., 1992. Non-traditional agricultural exports in Latin America. *Latin America Research Review,* 27 (2), 43-82.

Barrett, C.B., 2001. Measuring integration and efficiency in international agricultural markets. *Review of Agricultural Economics,* 23 (1), 19-32.

Beers, G., 2001. *Chain and network science: objectives, position and content.* KLICT, 's-Hertogenbosch.

Berdegué Sacristán, J.A., 2001. *Cooperating to compete: associative peasant business firms in Chile.* Proefschrift Wageningen

Bie Lilleør, H., Giné, X., Townsend, R.M., et al., 2005. *Weather insurance in semi-arid India: paper presented at annual bank conference on development economics, Amsterdam, May 23-24 2005.* [http://siteresources.worldbank.org/INTAMSTERDAM/Resources/GinePaper.pdf]

Bogetoft, P. and Olesen, H.B., 2004. *Design of production contracts: lessons from theory and agriculture.* Business School Press, Copenhagen.

Boselie, D., 2002. *Business case description: TOPS supply chain in Thailand.* Agrichain Competence Center/KLICT, Den Bosch.

Bowersox, D.J. and Closs, D.J., 1996. *Logistical management: the integrated supply chain process.* McGraw-Hill, New York.

Burt, R.S., 1997. The contingent value of social capital. *Administrative Science Quarterly,* 42, 339-365.

Burt, R.S., 2002. The social capital of structural holes. *In:* Guillen, M.F., Collins, R., England, P., et al. eds. *The new economic sociology: developments in an emerging field.* Russel Sage Foundation, New York, 148-192.

Coleman, J.S., 1990. *Foundations of social theory.* Belknap Press, Cambridge.

Cook, M.L. and Chaddad, F.R., 2000. Agroindustrialization of the global agrifood economy: bridging development economics and agribusiness research. *Agricultural Economics,* 23 (3), 207-218.

Cooper, M.C., Lambert, D.M. and Pagh, J.D., 1997. Supply chain management: more than a new name for logistics. *International Journal of Logistics Management,* 8 (1), 1-14.

Courtright, C., 2004. Which lessons are learned? best practices and World Bank rural telecommunication policy. *The Information Society,* 20 (5), 345-356.

Dirven, M., 1999. Dairy clusters in Latin America in the context of globalization. *International Food and Agribusiness Management Review,* 2 (3/4), 301-313. [http://www.ifama.org/nonmember/OpenIFAMR/Articles/v2i3-4/301-313.pdf]

Dolan, C. and Humphrey, J., 2000. Governance and trade in fresh vegetables: the impact of UK supermarkets on the African horticulture industry. *Journal of Development Studies,* 37 (2), 147-176.

Dolan, C., Humphrey, J. and Harris-Pascal, C., 1999. *Horticultural commodity chains: the impact of the UK market on the African fresh vegetables industry.* Institute of Development Studies, Brighton. Institute of Development Studies Working Paper no. 96.

Dorward, A., Kydd, J. and Poultron, C., 1998. *Smallholder cash crop production under market liberalisation: a new institutional economics perspective.* CAB International, Wallingford.

Dries, L. and Swinnen, J.F.M., 2004. Foreign direct investment, vertical integration, and local suppliers: evidence from the Polish dairy sector. *World Development,* 32 (9), 1525-1544.

Eisenhardt, K.M., 1989. Agency theory: an assessment and review. *Academy of Management Review,* 14 (1), 57-74.

Fafchamps, M., 2004. *Market institutions in Sub-Saharan Africa: theory and evidence.* Mit, Cambridge.

Farina, E.M.M.Q., 2002. Consolidation, multinationalisation and competition in Brazil: impacts on horticulture and dairy product systems. *Development Policy Review,* 20 (4), 441-457.

Farina, E.M.M.Q. and Reardon, T., 2000. Agrifood grades and standards in the extended Mercosur: their role in the changing agrifood system. *American Journal of Agricultural Economics,* 82 (5), 1170-1176.

Fearne, A. and Hughes, D., 1998. *Success factors in the fresh produce supply chain: some examples from the UK.* Wye College, London.

Fitter, R. and Kaplinsky, R., 2001. Who gains from product rents as the coffee market becomes more differentiated? A value chain analysis. *IDS Bulletin,* 32 (3), 69-82.

Gereffi, G., Humphrey, J. and Sturgeon, T., 2002. *Developing a theory of global value chains: a framework document: paper presented at global value chains conference, Rockport, Massachusetts, 25-28 April 2002.*

Gereffi, G. and Korzeniewitz, M., 1994. *Commodity chains and global capitalism.* Praeger, Westport.

Glover, D., 1990. Contract farming and outgrower schemes in East and Southern Africa. *Journal of Agricultural Economics,* 41 (3), 303-315.

Håkånsson, H., 1982. *International marketing and purchasing of industrial goods: an interaction approach.* Wiley, Chichester.

Håkånsson, H. and Snehota, I., 1995. *Developing relationships in business networks.* Routledge, London.

Harris-White, B., 1999. *Agricultural markets from theory to practice: field experiences in developing countries.* Macmillan Press, Basingstoke.

Hayami, Y. and Otsuka, K., 1993. *The economics of contract choice: an agrarian perspective.* Clarendon, Oxford.

Henson, S., Brouder, A.M. and Mitullah, W., 2000. Food safety requirements and food exports from developing countries: the case of fish exports from Kenya to the European Union. *American Journal of Agricultural Economics,* 82 (5), 1159-1169.

Henson, S. and Loader, R., 2001. Barriers to agricultural exports from developing countries: the role of sanitary and phytosanitary requirements. *World Development,* 29 (1), 85-102.

Hueth, B., Ligon, E., Wolf, S., et al., 1999. Incentive instruments in fruits and vegetables contracts: input control, monitoring, measurements and price risk. *Review of Agricultural Economics,* 21 (2), 374-389.

Humphrey, J. and Oetero, A., 2000. *Strategies for diversification and adding value to food exports: a value chain approach: UNCTAD conference on trade and development.* UNCTAD, Geneva.

Janssen, W.G. and Van Tilburg, A., 1997. Marketing analysis for agricultural development development: suggestions for a new research agenda. *In:* Wierenga, B., Tilburg, A.v., Grunert, K., et al. eds. *Agricultural marketing and consumer behavior in a changing world.* Kluwer, Boston, 57-74.

Jongen, W.M.F., 2000. Food supply chains: from productivity toward quality. *In:* Shewfelt, R.L. and Brückner, B. eds. *Fruit & vegetable quality: an integrated view.* Technomic, Lancaster, 3-18.

Kaplinsky, R., 2000. Globalisation and unequalisation: what can be learned from value chain analysis? *Journal of Development Studies,* 37 (2), 117-146.

Kaynak, H., 2003. The relationship between total quality management practices and their effects of firm performance. *Journal of Operations Management,* 21 (4), 405-435.

Keesing, D.B. and Lall, S., 1992. Marketing manufactured exports from developing countries: learning sequences and public support. *In:* Helleiner, G.A. ed. *Trade policy, industrialization and development: new perspectives: study prepared for the World Institute for Development Economics Research (WIDER) of the United Nations University.* Clarendon, Oxford, 176-193.

Key, N. and Runsten, D., 1999. Contract farming, smallholders, and rural development in Latin America: the organization of agroprocessing firms and the scale of outgrower production. *World Development,* 27 (2), 381-401.

Krueger, A.O., Schiff, M. and Valdes, A., 1988. Agricultural incentives in developing countries: measuring the effect of sectoral and economywide policies. *World Bank Economic Review,* 2 (3), 255-271.

Kumar, A., Margulis, S.T. and Motwani, J., 2001. An efficient real-world food delivery system: the Dabbawallas of Mumbai. *IIMB Management Review,* 13 (4). [http://www.iimb.ernet.in:8080/review/abs134.htm]

Kuyvenhoven, A. and Bigman, D., 2001. *Technical standards in a liberalised agri-food system: institutional implications for developing countries: paper presented at workshop on capacity building in developing countries regarding non-tariff barriers to trade (SPS), The Hague, June 20, 2001.*

Lancioni, R., Smith, M. and Oliva, T., 2000. The role of the Internet in supply chain management. *Industrial Marketing Management,* 29 (1), 45-56.

Lazzarini, S.G., Chaddad, F.R. and Cook, M.L., 2001. Integrating supply chain and network analyses: the study of netchains. *Journal on Chain and Network Science,* 1 (1), 7-22.

Luning, P.A., Marcelis, W.J. and Jongen, W.M.F., 2002. *Food quality management: a techno-managerial approach.* Wageningen Pers, Wageningen.

Marsden, T. and Wrigley, N., 1996. Retailing, the food system and the regulatory state. *In:* Wrigley, N. and Lowe, N. eds. *Retailing, consumption and capital: towards the new retail geography.* Longman, Harlow, 33-47.

Omta, O., 2004. Management of innovation in chains and networks. *In:* Camps, T., Schippers, A. and Hendrikse, G. eds. *The emerging world of chains and networks: bridging theory and practice.* Reed Business Information, 's-Gravenhage, 205-218.

Omta, S., Trienekens, J. and Beers, G., 2001. Chain and network science: a research framework. *Journal on Chain and Network Science,* 1 (1), 1-6.

Otsuki, T., Wilson, J.S. and Sewadeh, M., 2001. Saving two in a billion: quantifying the trade effect of European Food Safety Standards on African exports. *Food Policy,* 26 (5), 495-514.

Porter, M.E., 1990. *The competitive advantage of nations.* MacMillan, Basingstoke.

Porter, M.E., 1998. Cluster and the new economics of competition. *Harvard Business Review,* 76 (6), 77-90.

Porter, M.E., 2001. Strategy and the internet. *Harvard Business Review,* 79 (3), 62-78.

Raikes, P., Jensen, M.F. and Ponte, S., 2000. Global commodity chains analysis and the French filière approach: comparison and critique. *Economy and Society,* 29 (3), 390-417.

Reardon, T. and Barrett, C.B., 2000. Agroindustrialization, globalization, and international development: an overview of issues, patterns, and determinants. *Agricultural Economics,* 23 (3), 195-205.

Reardon, T., Codron, J-M., Busch, L., et al., 1999. Global change in agrifood grades and standards: agribusiness strategic responses in developing countries. *International Food and Agribusiness Management Review,* 2 (3/4), 421-435. [http://www.ifama.org/nonmember/OpenIFAMR/Articles/v2i3-4/421-435.pdf]

Reardon, T. and Timmer, C.P., in press. Transformation of markets for agricultural output in developing countries since 1950: how has thinking changed? *In:* Evenson, R.E., Pingali, P. and Schultz, T.P. eds. *Handbook of agricultural economics. Vol. 3: Agricultural development: farmers, farm production and farm markets.* Elsevier, Amsterdam.

Regmi, A. and Gehlhar, M., 2003. Consumer preferences and concerns shape global food trade. *FoodReview,* 24 (3), 2-8. [http://www.ers.usda.gov/publications/FoodReview/septdec01/FRv24i3a.pdf]

Rindfleisch, A. and Heide, J.B., 1997. Transaction cost analysis: past, present, and future applications. *Journal of Marketing,* 61 (4), 30-54.

Ruben, R. and Lerman, Z., 2005. Why Nicaraguan peasants stay in agricultural production cooperatives? *European Review of Latin American and Caribbean Studies* (78), 31-48.

Ruben, R., Saenz, F. and Zuniga, G., 2004. Contracts or rules: quality surveillance in Costa Rican mango exports. *In:* Hofstede, G.J., Schepers, H., Spaans-Dijkstra, L., et al. eds. *Hide or confide? the dilemma of transparency.* Reed Business Information, 's-Gravenhage, 51-58.

Saenz, F. and Ruben, R., 2004. Export contracts for non-traditional products: Chayote from Costa Rica. *Journal on Chain and Network Science,* 4 (2), 139-150.

Scott, G.J., 1995. *Prices, products and people: analyzing agricultural markets in developing countries.* Rienner, Boulder.

Sheldon, I.M., 1996. Contracting, imperfect information, and the food system. *Review of Agricultural Economics,* 18 (1), 7-19.

Sturgeon, T.J., 2001. How do we define value chains and production networks. *IDS Bulletin,* 32 (3), 9-19. [http://www.ids.ac.uk/globalvaluechains/publications/Sturgeon.pdf]

Trienekens, J.H., 1999. *Management of processes in chains: a research framework.* Proefschrift Wageningen

Trienekens, J.H. and Hvolby, H.H., 2001. Models for supply chain reengineering. *Production Planning and Control,* 12 (3), 254-264.

Uzzi, B., 1997. Social structure and competition in inter-firm networks: the paradox of embeddedness. *Administrative Science Quarterly,* 42, 35-67.

Van der Laan, H.L., 1993. Boosting agricultural exports: a marketing channel perspective on an African dilemma. *African Affairs* (92), 173-201.

Van der Laan, H.L., Dijkstra, T. and Van Tilburg, A., 1999. *Agricultural marketing in tropical Africa: contributions from the Netherlands.* Ashgate, Aldershot.

Van der Vorst, J.G.A.J., 2000. *Effective food supply chains: generating, modelling and evaluating supply chain scenarios.* Proefschrift Wageningen [http://www.library.wur.nl/wda/dissertations/dis2841.pdf]

Van Tilburg, A. and Moll, H.A.J., 2000. *Agricultural markets beyond liberalization.* Kluwer Academic Publishers, Boston. Selected contributions of the international seminar 'Agricultural markets beyond liberalization', held at Wageningen University in September 1998

Vellema, S. and Boselie, D., 2003. *Cooperation and competence in global food chains: perspectives on food quality and safety.* Shaker Publishing, Maastricht.

Weaver, R.D. and Kim, T., 2001. *Contracting for quality in supply chains: paper presented at 78th EAAE Seminar Economics of contracts in agriculture and the food supply chain, Copenhagen, June 2001.*

Williamson, O.E., 1987. *The economic institutions of capitalism: firms, markets, relational contracting.* Free Press, New York.

Williamson, O.E., 1999. Strategy research: governance and competence perspectives. *Strategic Management Journal,* 20 (12), 1087-1108.

CHAIN INTEGRATION
AND DEVELOPMENT

CHAPTER 2

AGRICULTURAL DEVELOPMENT AND TRADE LIBERALIZATION

ROBERTO RODRIGUEZ

Minister of Agriculture, Brazil

Abstract. Agriculture represents in Brazil roughly one third of GDP; a quarter of employment and 42% of exports. During the last 15 years, the cultivated area increased with 24%, but production more than doubled. Much attention is given to innovation in new technologies, fertilizers and agrochemicals for enhancing productivity. Supply in the agribusiness sector is continuously increasing, and Brazil has become a major exporter of coffee, sugar, soybeans and meat to the EU, US and Asian markets. Even while there is still some room for expansion, it is recognized that most of future agricultural growth must come from productivity improvement.

Open markets are in the benefit of their consumers, but countries are entitled to maintain some subsidies for environmental and social purposes as long as these do not generate distortions in the market. This is the key question that has to be faced in the WTO negotiations. The core challenge for Brazil is to make agro-food chains a development instrument for the government. Cooperatives play a very important role in integrating chains and enhancing a participatory process of rural development. They bring small farmers together, adding value to their production and enabling them to access the market.

Keywords: agricultural growth; agricultural productivity; market liberalization; cooperatives; Brazil

INTRODUCTION

We owe much to Wageningen University for this opportunity to discuss some key issues related to agriculture and trade in the world. I structured my presentation in two parts. First, I will show you how important the agricultural sector is for a country like Brazil. Second, I am going to discuss the issue of globalization and trade intervention and why these are critical for countries like Brazil. Before this, I would like to share with you some of my personal experiences and how I learned about the importance of international economic relations.

FARMERS AND STOCK EXCHANGE

My grandfather, the father of my father, was a large coffee farmer in Brazil. In the beginning of the 20th century he owned about seven coffee estates. He was a very severe, but also very rich man in Sao Paulo state. Being an important farmer, he was

29

R. Ruben, M. Slingerland and H. Nijhoff (eds.), Agro-food Chains and Networks for Development, 29-39.
© 2006 Springer. Printed in the Netherlands.

also involved in politics and finally he even became the mayor in his city. He owned a nice car and a good house, and became engaged to a beautiful girl. At that time he was a symbol of wealth. Soon, however, he experienced the breakdown of the stock exchange in New York in 1929. My grandfather had never heard about New York and did not know anything about the stock exchange. But soon he found out that the pieces of land he owned only had 10-20% of the value they used to have. He had just bought a new farm. At that time he used to pay 10% of the price and the rest would be paid during the next 4-6 years. But after 1929, he noticed that the purchase price of his last acquired farm was more or less the same as the value of all the other seven estates together. He therefore had to give the seven estates as collateral for his loan to the Banco de Brazil, our official bank. That is how he became poor again and also ashamed, and decided to go away from Piricicaba, his native city. He became engaged in cutting wood and suffered from such a poor situation that when he died and was buried, his brothers and sons, once informed, could not even find his grave.

So, a rich man can become poor just because of a New York bankruptcy. But, the father of my mother was an Italian immigrant. He went to Brazil and became an employee of a coffee farmer who experienced a bankruptcy in 1929 and had to give his farm to the Bank of Brazil. The bank sold this land to the employees of the farms. The father of my mother became a poor family farmer, due to the same reason that the father of my father, who once was a rich farmer, ended up poor. The interesting thing is that both did not know anything about New York and had nothing to do with the stock exchange, but were unwillingly affected by these affairs. For some it meant bankruptcy, for others it was a change. My own father was a clever man, supported by my mother, and I became finally the Minister of Agriculture of Brazil. This history orients me in my daily work as a Minister. We have the responsibility to provide the conditions for farmers in order to avoid that some external events put them into misery, while at the same time we should offer them the opportunities for reaping the fruits of progress.

THE ROLE OF AGRICULTURE

Let me now show you in a glimpse how the Brazilian agri-business sector is structured. Agriculture represents roughly one third of our GDP; 27% of our jobs come from the agri-business sector and agricultural products represent 42% of our exports. During the last 15 years, the cultivated area has increased with 24%. During the same period, production has increased with 105% (see Table 1). This year, 2004, we could have had even more production, but due to a terribly dry season in the south of the country large areas were lost. Nevertheless, you can notice that farmers have put a lot of effort in increasing their land productivity, making use of the innovations generated by EMBRAPA (the Brazilian research and extension agency).

Table 1. *Evolution of Brazilian Grain Crops*

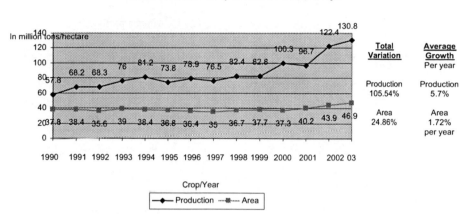

Source – CONAB/MAPA

The same is happening for different kinds of beef, pork and chicken meat (see Table 2). Particularly through the development of specific product labels (e.g. red lion for beef, blue-eye chicken, and green lion for pork meat), all the three sectors have reached a substantial increase in production and productivity.

Table 2. *Brazilian Meat Procuction*

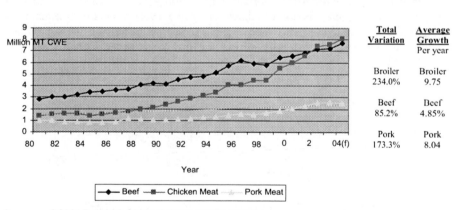

Source – CONAB (1995-2003) & FAO (1980-1984)
Elaboration - ICONE
Note: 2003/04 forecast

Agricultural growth is strongly favoured by the use of new technologies, fertilizers and agrochemicals. It is obvious that – as population is increasing in the whole world – the available farmable area per capita will be reduced. So in the future, there will be land scarcity. The core question is then: will there be enough land to feed

Table 3. *Agricultural Intensification: Fertilizers and Agrochemicals Sales*

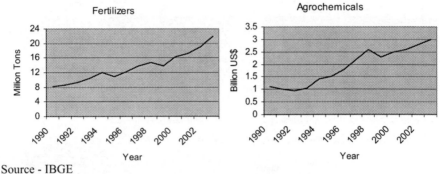

Source - IBGE
Elaboration – ICONE

mankind? Wide availability of modern agricultural technologies for improving yields is therefore of fundamental importance. Over the past 12 years, we can notice that the use of fertilizers and agrochemicals has more than doubled in Brazil (see Table 3), in an effort to increase agricultural productivity.

EXPORT PERFORMANCE

This productivity growth has important implications for our trade performance. There has always been a positive balance since the 1960s and this is maintained during the 1970s and the 1980s (see Table 4). Supply in agri-business has been continuously increasing, especially during the last three years. For this year we

Table 4. *Brazilian Agribusiness Trade Balance*

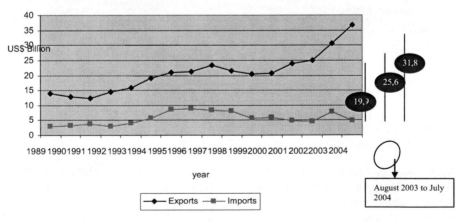

Fonte – MAPA/SPC

expect to reach something close to 32 billion dollars of surplus in the sector. The European Union is the most important market for Brazil: one third of our exports in the agri-business sector are directed to the European Union, followed by the United States with 14%, and Asia with a share close to 20%, although the latter is rapidly increasing due to the expansion of the Chinese market.

Table 5. Brazil's shares in world production and trade (2004)

	Production		**Exports**	
	Share in world production	Ranking	Share in world production	Ranking
Coffee	31 %	1	29 %	1
Orange juice	47 %	1	82 %	1
Sugar	16 %	1	29 %	1
Soybeans	30 %	2	38 %	1
Soybean starch	18 %	2	34 %	2
Soybean oil	19 %	2	28 %	2
Coffee (soluble)	n.a.	n.a.	44 %	1
Poultry	14 %	3	29 %	2
Beef	16 %	2	20 %	1
Tabacco	9 %	3	23 %	1
Cotton	5 %	5	5 %	4

Fonte – ERS/USDA
Elaboration - ICONE

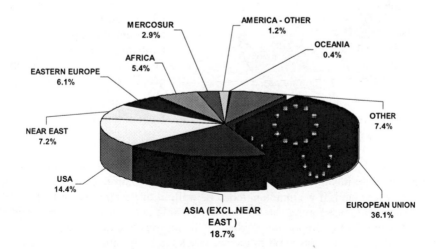

Brazil is one of the most important producers of coffee, orange juice and sugar, and the world's leading exporter of soybean, beef and tobacco (see Table 5). We are becoming important exporters of cotton and bio-fuels as well. Recently, during one of my visits to Asia, I became involved in a lengthy discussion about bio-fuels. Countries like Thailand do not possess any oil and fully depend on the imports of oil at high prices. For these countries, further development of biological sources for energy is of vital importance. We should be aware of the fact that in less than one century, humanity became fully dependent of a product that is going to finish some day. This should be a collective concern of mankind. We now depend on six or seven large international companies and if we want to escape from this trap, the further development of bio-fuels might be the right alternative.

LAND USE

In addition to cropping, Brazil plays an important role in livestock production. We have 62 million hectares occupied by the agricultural sector, of which 47 million are with food crops and 15 million with cash crops (see Table 6). In addition, we cultivate 220 million hectares of pastures. Some research organized by the fertilizers sector in Brazil informs us that in the next 15 years about 30 million hectares of

Table 6. Land-use distribution in Brazil (by categories and sectors; 2004)

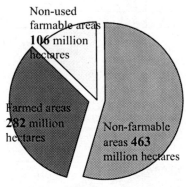

Brazil Land Use - Actual and Potential

Territorial distribution (million ha)	
Amazon rainforest	350
Breeding pastures	220
Protected areas	55
Annual cultures	47
Permanent cultures	15
Cities, towns, lakes, roads and swamps	20
Cultivated forests	5
Other uses	38
Total	707
Other usus	38
Unexploited area	106

Source – VEJA 03032004
Source – IBGE e CONAB - MAPA

pastures will be transformed into agricultural areas because the technological progress in cattle-raising is so large that we will be able to produce much more beef while using less land. If we maintain today 62 million of hectares and it took us 500 years to arrive at these figures, how should we be able in the next 15 years to add 30 million hectares? More importantly, what are the reasons to expect such changes? A lot of this discussion is based on future scenarios made by international enterprises and the food industry. They have – so to say – discovered that Brazil has to produce 60 million tons of grain more in the next ten years. Both studies conclude that we might have to incorporate into agriculture in the next decades a total area of two million hectares per year. This is something that looks very difficult and may be impossible. During the last three years we have been incorporating 2.5 million hectares per year, and last year we reached 3.5 million hectares just in one year, mainly through conversion of pastures into arable land.

It is highly important for us to understand what is going to happen with the agricultural sector of Brazil in the near future. Different media have been speaking about the expansion of pastures in the Amazon region and their disappearance from the region of São Paulo state, where land has become too expensive to maintain extensive pasture areas. In other regions, we need to develop agriculture, given the large potential for Brazilian agri-business and its predominant role in our economy, contributing roughly one third of the GDP, one third of the jobs and almost half of exports. Even while agriculture is now already the largest sector in the country, we still have an important potential to grow, perhaps with 50% or more during the next decade. The largest potential is available in the so-called *cerrados*, a kind of savannahs in Brazil where we have 90 million hectares of suitable agricultural area. The *cerrados* used to be very poor areas, but with potential to be developed for cropping purposes. There are no great fertility problems, and mechanization can be easily implemented. In Mato Groso state, we are now harvesting soybean and seeding corn for the next season. These areas possess appropriate conditions to convert this potential into reality. But at the same time we need to recognize that at least 85% of future agricultural growth should come from productivity improvements of land already in production. The foreseen increase in population and the reduction of arable farmland (as illustrated in Table 7) asks for substantial increases in productivity.

Table 7. World Agriculture Area Per Capita

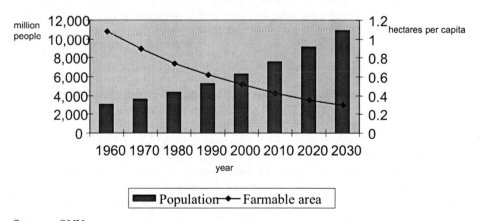

Source – ONU
OBS: Total Area Used = 3,234,521 ha

On the other hand, developing this region would enable us not to touch land in the Amazon region. What is happening in the Amazon region is that the controls that are implemented today in order to avoid deforestation are very useful and become very efficient, particularly regarding the larger farmers whose activities can be controlled through satellites and other methods (Cattaneo 2002). But millions of (very) small farmers just cut a few trees every year and this can impossibly be controlled. They need first of all access to other income sources or better-rewarding activities in order to prevent the degradation of natural resources.

FUTURE FOR SMALLHOLDERS

This brings us to one of the most important problems we are facing today. While we possess enough land and suitable tropical technologies for the agricultural sector and we have EMBRAPA as our federal organization for research and extension (maintaining partnership with 82 different countries in the world), a wide range of state organizations, universities and even some private organizations, especially in the cooperative sector, that are organizing excellent research for new technologies, and especially we have good farmers, still the progress reached in productivity development of the smallholder sector or so-called family agriculture during the last four decades has been rather limited (Cassel and Patel 2003).

It is important to recognize that we have more and younger farmers, as well as a large number of female farmers in Brazil involved in agriculture. In some European countries, women do not want to get married anymore with farmers, and revenues from farming in developing countries are reduced because of ritual inheritance regulations. We need therefore to guarantee that farmers perceive profits and can establish decent families in the countryside. When I am travelling throughout my country during the weekends, visiting fairs in different regions of the country, I am always delighted to meet with young people who received a good training in agronomy or rural-economic studies. Our rural population thus needs further education and qualified people prepared to meet the future challenges.

GLOBALIZATION AND TRADE LIBERALIZATION

I have been watching what happened in the world after the globalization of the economy (Farina and Viegas 2003). Globalization has two sides; first we see a positive impact because trade has been growing, and if there is more trade there will be more production and more wealth throughout the world. But we can also notice a downside of globalization. The negative effects of globalization are the increasing social exclusion and the overwhelming concentration of wealth in all countries in the world. Wherever you go, you see social exclusion and concentration of wealth, especially and mostly in developing countries. Nowadays, these two faces of the process of globalization are more and more threatening democracy and peace. In the newspapers and other media, we notice that democracy is at risk and that peace is even more threatened. Therefore, the most important challenge for mankind in the 21st century is maintaining democracy and peace. To do so, it is absolutely fundamental to reduce the social and economic gaps all over the world, between rich and poor countries, and between rich and poor people within the same country. This is the most important challenge for human development in our times.

While everybody agrees on the necessity to redistribute income, it is always another's income, not mine. Well-trained academic professionals teach us that the easiest way to reduce this gap is when the developed countries open up their agricultural markets for products from the developing countries. First, because in the developed countries there is only a small minority of the population depending on farming, while in the developing world a large number of people depend on agricultural income. Second, richer countries can afford to pay their farmers for not producing, while developing countries have to produce to pay the debts. I strongly believe that agricultural markets have to be opened in order to defend democracy and peace, and to reduce the existing development gaps. I am also convinced that this is going to happen, because governments in developed countries will recognize – even if they face some electoral problems – that open markets are in the benefit of their consumers. This does not mean, however, that all subsidies fully need to disappear. Many countries are entitled to maintain subsidies to stimulate multifunctional land use, to defend the environment and to foster sustainable agriculture, and even to maintain people living in the rural areas for reasons of social and political stability. Otherwise, subsidies should not generate externalities to other countries, disturbing the price at the market or limiting the possibilities for developing countries to accede these markets. Subsidies may be necessary, but should not generate distortions in the market. This is the key question that has to be faced in the WTO negotiations.

The new WTO agreement reached on 1 August 2004 had three main pillars: market access, domestic support and price subsidies. Tariff reductions for the future still do not imply a real opening of markets. It is just a first sign of opening, but it is certainly not a guarantee. We need much more than that, in fact we need bilateral agreements. We shall forward our case in the good relationship between the European Union and MERCASUR to get clear and immediate market access for Brazil, the Latin-American countries and the developing world in order to be able to generate our own wealth and to reduce the enormous social and economic gaps. This permits us in turn to defend democracy and peace.

CHAIN DEVELOPMENT

To reap the advantages of trade, we trust on the generation of specialized and qualified young farmers throughout the country. The core challenge we face in Brazil is how to make agro-food chains a development instrument for the government. Sustainability has become a concept that everybody is talking about. We are doing our job in promoting organic and minimum-tillage agriculture and integrated cropping systems. We are now improving sanitary measures and are strongly involved in international negotiations on export promotion, leading the DG20 group that is responsible for the success of the last round of negotiations in WTO.

Chain coordination is something we are trying to promote, especially from academic circles, orientating the actors and the market agents to establish their own governance in order to become less dependent on governmental decisions. The establishment of chain organizations is based on sectoral chambers that represent all stakeholders within particular production chains. Today we have 21 chambers working very successfully since we put together scientists, the producers of inputs, the cooperatives, the primary producers, the exporters, the industries and the distribution sector. Everybody is discussing about integral chain performance in order to improve sustainability of the chain to satisfy consumers (Zylbersztajn and Filho 2003; Neves et al. 2000). The price and quality of food that the consumers need today drive us into the right direction. One of the chambers is dealing with the chicken, pork and corn chain. People came to me and asked why chicken, pork and corn altogether? I explained to them that a chicken is nothing else than an egg full of corn, with wings and backs. If you do not have eggs and corn you cannot produce chicken. They have to understand that they must work together in order to get sustainability in the chain.

Today infrastructure and logistics represent in Brazil the most critical bottlenecks for chain integration. During the last 10 years we have not made sufficient investments in infrastructure, ports and railways, and this is now becoming a large problem for us. Lulu's government is trying to promote in parliament that public–private partnership can become an attractive device for infrastructure development. But this requires new legislation that we will have to attract private investors for co-investment in infrastructure work.

PARTNERSHIPS AND COOPERATION

Supply-chain partnerships are especially important to us, particularly to support the economic transition from intensive use of the environment towards less-intensive but more efficient production systems. Cooperatives are playing a very important role in integrating chains both locally and worldwide. One of the questions for this conference was: how to make income available to farmers that participate in the chain in order to enhance a participatory process of rural development. From 1997 to 2001, I have been president of the international cooperative alliance (ICA), an important organization with its head-office in Geneva. Before that, I have been president of the Brazilian federal cooperative organization and the cooperative organization for the Americas, strongly committed to the support for agricultural cooperatives. For more than 10 years I have been working in the international cooperative movement. During the last 15 years I have been travelling to visit cooperatives all over the world, in more than 80 countries. I learned that cooperatives can provide a very sound answer to the question of participation in income sharing. Cooperatives may be the only way to bring small farmers together, adding value to their production and enabling them to access the market.

REFERENCES AND FURTHER READING

Cassel, A. and Patel, R., 2003. *Agricultural trade liberalization and Brazil's rural poor: consolidating inequality.* Food First/Institute for Food and Development Policy, Oakland. Policy Brief no. 8. [http://www.globalpolicy.org/globaliz/econ/2003/08agribrazil.pdf]

Cattaneo, A., 2002. *Balancing agricultural development and deforestation in the Brazilian Amazon.* IFPRI, Washington. Research Report no. 128. [http://www.ifpri.org/pubs/abstract/129/rr129toc.pdf]

Farina, E.M.M.Q. and Viegas, C.A.d.S., 2003. *Multinational firms in the Brazilian food industry: symposium paper presented at IFAMA World Food & Agribusiness, Cancun.* [http://www.ifama.org/conferences/2003Conference/papers/EFarina.pdf]

Ministerio da Agricultura Pecuária e Abastecimento, 2004. *EMBRAPA: Empresa Brasileira de Pesquisa Agropecuária.* Available: [http://www.embrapa.br/] (2005).

Neves, M.F., Chaddad, F.R. and Lazzarini, S.G., 2000. *Alimentos: novos tempos e conceitos na gestão de negócios.* Editora Pioneira, Sao Paolo.

Zylbersztajn, D. and Filho, C.A.P.M., 2003. Competitiveness of meat agri-food chain in Brazil. *Supply Chain Management,* 8 (2), 155-165.

CHAPTER 3

PARTICIPATION OF SMALLHOLDERS IN INTERNATIONAL TRADE

LEONARD NDUATI KARIUKI

National Chairman
Kenya National Federation of Agricultural Producers (KENFAP)
P.O. Box 43148-00100, Nairobi, Kenya
E-mail: farmers@knfu.org

Abstract. In many developing countries, the production and trade in agricultural products plays a crucial role for economic growth and development. It enables countries to earn foreign exchange, to create employment and to invest in sustainable utilization of resources. However, developing countries are often faced with a number of constraints that limit local farmers to develop. First of all, the agricultural sector often remains underdeveloped as a result of inadequate and unbalanced investments compared to the industrial sector. At the same time, farmers have insufficient access to market information and (as a result) have poor bargaining power. Most farmers own small plots of land, and as a result are unable to invest in meeting the stringent international standards to which they need comply. Finally, it is felt that, also at WTO level, trade rules favour multinationals and the developed countries, and not small farmers in developing countries. For these farmers to participate in international trade, KENFAP identified three strategies. The first is to build institutional capacity and self-organization; these will enable farmers to be better informed, improve bargaining power and learn from international best practices. The development of partnerships and networks will further contribute to preparing for market access. Finally it recommends the active support through enabling incentives (policies and regulations) from national and international policymakers.
Keywords: international trade; food safety; bargaining power; partnerships

THE IMPORTANCE OF AGRICULTURE IN DEVELOPING COUNTRIES' ECONOMIES

The agricultural sector plays a crucial role in most of the developing countries' economies. In Kenya's economy for instance, agriculture is still the dominant sector. The sector contributes 25% of GDP directly to the economy and is estimated to contribute another 27% indirectly through linkages with other sectors. It also contributes about 45% of government recurrent revenue and 75% of industrial raw materials, while contributing employment opportunities to about 77% of the population.

R. Ruben, M. Slingerland and H. Nijhoff (eds.), Agro-food Chains and Networks for
Development, 41-48.

From a gender perspective, women dominate agricultural activities and therefore any effects on the agricultural sector would have direct impact on livelihood of women and children who depend on them. In Kenya, there is a direct positive relationship between growth in the agricultural sector and that of the entire economy.

Figure 1 illustrates that whenever the agricultural sector has performed well, the national economy also performs well and vice versa. Key factors that affect growth of the agricultural sector will thus affect the growth of the overall economy and would have an impact on the poverty levels in the country.

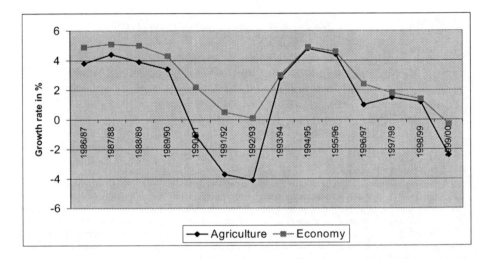

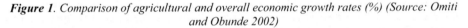

Figure 1. Comparison of agricultural and overall economic growth rates (%) (Source: Omiti and Obunde 2002)

Trade at local and international level plays a crucial role in economic development of a country. Many countries vigorously pursue international trade opportunities, since this is the best way to economic development. Indeed, some Southeast Asian countries that embraced free-market policies early have made great strides in economic development, although state controls and interventions backed these resounding successes.

Most countries of the world have implemented a series of economic reform measures since the mid 1980's. Sweeping reforms were undertaken in the early 1990's aimed at encouraging participation of the private sector in production, marketing, processing and trading of agricultural commodities.

THE ROLE OF TRADE IN POVERTY ALLEVIATION IN DEVELOPING COUNTRIES

The five major roles of trade in reducing poverty in developing countries are the relation to: economic development; food security and sovereignty, foreign-exchange earnings; creation of employment; and effective utilization of resources.

Trade and economic development

Countries worldwide have achieved various levels of development through various forms of trade, both internal and international. Trade at local and international level plays a crucial role in income generation, which could be ploughed back into various forms of development initiatives, aimed at wealth creation and therefore poverty alleviation. Many countries vigorously pursue international trade opportunities, since the gains through this form of exchange lead to faster wealth generation and therefore increased income to the players. This turns out to be the best way or strategy to faster economic development.

Trade and food security / sovereignty

Productivity in developing countries will continue to be low as long as the producers do not have value for their product. Food security will be an issue of the developing world so long as the products do not access good markets in terms as of their value. It will remain increasingly difficult to enhance sufficiency in food production if there are no markets for the produced goods.

The achievement of food security has however been misinterpreted in certain circles as producing enough food to be self-sufficient. It has been argued that once a country is able to produce enough food for its own use it can then embark on the next step of producing for export or industrialization. This belief has prompted debate of whether developing countries should not emphasize on food production for food self-sufficiency instead of production for export. Such are the paradoxes abound in the developing world.

This theory may be true only under certain circumstances, which may include:
- export earns too little to be able to support importation of food
- a country has other trading resources, e.g., petroleum, minerals or service industries.

All this is however possible where there is vigorous trade that allows the farmers to market their produce easily and profitably and allows them to obtain their other requirements such as inputs, personal and other support requirements including food. The contra-argument is that a country needs not to produce the food it requires to be food-secure.

Trade allows a country that produces little of its staple food to maximize on producing whatever it does best, and use the money to buy the food it requires. The prime consideration in this case is: where does a country or a region have a relative advantage? Is it in producing food for consumption or producing for export? Trade therefore gives countries or regions the right to choose, which is generally food security as it is regarded as food sovereignty.

Trade and foreign-exchange earning

In most developing countries international agricultural trade is the major source of foreign exchange. For instance, over the year 2003, Kenya exported 61,000 tons of cut flowers, over 23,675 tons of fruits and 48,674 tons of fresh vegetables, capturing a total of export revenue form horticulture of over US $ 350 million. This is about 10% of the total national budget for one year. In nominal terms, Kenya has more than quadrupled export revenues from horticulture since 1995.

Trade and employment creation

Trade allows for establishment of industries at local levels either on the basis of on-farm processing, small-scale rural industries or medium-scale agro-processing. These industries offer employment to the communities. The industries act as outlets for increased agricultural products, therefore encouraging uptake of farming as a livelihood support mechanism and therefore on-farm employment. In most developing countries over 70% of the population is engaged in one form of agricultural-related employment or the other. A vibrant farming sector is one backed up by a reliable market, a situation that encourages more entrants into production.

Trade and effective utilization of resources

Trade enables the small-scale producers to utilize the locally available resources such as land effectively. Agricultural trade is essential for the conservation of the world's natural resources such as farmland, forests and water for economic development of rural communities. Such utilization is indirectly fuelled through increased value for the resources, out of the value for the products. In retrospect, a lack of trade leads to redundancy and underutilization of available land and related resources.

CONSTRAINTS TO INTERNATIONAL TRADE BY SMALLHOLDERS FROM DEVELOPING COUNTRIES

We can identify six types of constraints that limit smallholders to enter international markets. These are an underdeveloped sector, insufficient market information, poor bargaining power, stringent market conditions, WTO regulations, and a lack of institutional capacity.

Underdeveloped agricultural sector

The underdevelopment of an agricultural sector is often the result of inadequate allocation of resources to this sector. In Kenya for example, 80% of the population live in rural areas with 60% of this population living below the poverty line. However, traditionally there has been preference for an industrial and manufacturing sector from the fiscal resource allocation, as opposed to the agricultural sector, especially during the import-substitution policy regime. The rural areas are still predominantly non-commercial and budgetary allocations are a testimony show for bias against agriculture by the government. This trend has consistently fostered marginalizing farmers, leading to low participation in trade both at domestic and international levels.

Inadequate investment in all-round agricultural activities is a major cause of the underdeveloped agricultural sector. The investors tend to focus on the commodities doing well in the international market at the nick of the moment, hence ignoring the other enterprises. Farmers under the circumstances therefore cannot participate in international trade. In most of the developing countries, there has been over-reliance on supply-driven services. Producers do not demand for such services as extension and market information.

Insufficient market information to farmers

Market information is the key to influencing the decisions at the farm level as concerns the choice of enterprise. Most smallholder farmers in developing countries lack timely, relevant and reliable information about the market possibilities that exist including:
- type and quality of product demanded
- market regulations
- seasons of demands and fluctuations
- price or price fluctuations.

Poor bargaining power

Due to their weak positioning at the international market place, smallholder farmers in developing countries almost always suffer poor bargaining power. This is occasioned by inadequate involvement of farmers in the decision-making processes. Farmers in developing countries have borne the full brunt of economic reforms beginning with structural adjustment programmes that culminated into liberalization and globalization, whose climax was the signing of the World Trade Organization's agreements. The direct effects of these on farmers are the collapse of marketing boards and cooperative societies that used to cushion farmers against exploitation. Smallholder farmers therefore have consequently become price takers.

Stringent market conditions

Most farmers in developing countries fail to meet the required standards on production and primary processing, which involves packaging, labelling and transportation of products. The farmers do not have the opportunities for harmonizing existing local standards with required international market standards. Such standards include:

- Sanitary and Phytosanitary Standards (SPS)
- European Retailers Code of Good Agricultural Practices (EUREPGAP), which demands traceability of produce from the retail shelf back to the farm gate through a complicated and costly certification proc ss by accredited companies.

For instance, while some of EUREPGAP provisions constitute a progressive approach that will in the long run contribute to upgrading the supply chain of produce exported by developing countries, some of the provisions are not realistic enough with respect to local conditions. Smallholder farmers in many developing countries have very small parcels of land ranging from half an acre to 2 acres. It would easily require such a farmer 2 to 3 years of production to pay for one annual EUREPGAP audit, making participation of smallholders simply impossible.

WTO regulations

Rules of agricultural trade create imbalances that favour multinationals and developed countries. Key elements here include:

- export subsidization which is clearly trade distorting.
- production subsidies through domestic support, which are by and large also trade distorting
- import barriers, e.g. EUREPGAP and SPS.

Lack of institutional capacity

There are three main institutional elements that hamper inclusion of smallholders in international trade: contract farming; underdeveloped entrepreneurial skills; and the inability of farmers to influence policy-making.

- Contract farming: smallholder farmers in developing countries are inadequately supported by organizations (e.g. NGOs), leading to their weak capacity and knowledge on contracts. Some farmers sell to exporters under stringent contracts and depend on the exporters' willingness to share information on the demand in international markets. They lack information such as: who is willing and able to enter into contract; repercussions of failure to adhere to the contract; and occasionally premature withdrawal of contactors.
- Low entrepreneurial skills: the majority of smallholder producers in the developing world practise farming as a way of life and not necessarily as a business. Even where they have been given technical skills and knowledge to enhance profitable productivity, no business training is given for them to

transform such skills into businesses. There is little knowledge of matters such as production planning, marketing, contracts, business units, financing and bookkeeping, and negotiating skills.
- Regulatory framework and inability of farmers to influence policy: in many African countries farmers form the majority of the population and in many instances contribute the bulk of national wealth, yet they are the least vocal and least influential when it comes to policy formulation on matters affecting them. This is due to lack of organized forums, and where they exist as individuals, they are terribly weak and lack capacity to voice their concerns effectively.

STRATEGIES AND INCENTIVES TO GUARANTEE BETTER PARTICIPATION IN INTERNATIONAL TRADE

The three main strategies and incentives that have been identified to guarantee better participation of smallholders in international trade are: building institutional capacity; establishing local, national and international networks; and long-term policies and regulations.

Building institutional capacity

Capacity building among farmers through training in entrepreneurial skills is believed to be an important tool towards revitalizing the lost power. This will enhance awareness on the potential that exists in agri-business, realizing and appreciating the problems in achieving this potential and prepare them to be an effective force for lobbying and advocacy on issues that contribute to the problems.

Producers need to be facilitated in self-organization beyond out-grower schemes to achieve an independent professional organization at national level. Such an organization would help in giving smallholder farmers the capacity to:
- access information on markets
- negotiate prices
- learn from international best practices.

Creating partnerships

Partnerships and networks can be created at local, national and international level.

Local networks

At local levels there are persons and institutions that separately hold assets and skills that individually can only lay idle while, when being combined and made complimentary to one another, they can form viable production and marketing units for the benefit of all contributors:
- the land owner, with idle or underutilized land
- the agricultural graduate, unemployed but a potential resource person
- the local marketing expert, to organize marketing strategies

- the transporter, with an idle transport vehicle
- the asset owner, with idle agricultural machinery.

Partnerships between these parties could create a formidable force. However, training and motivation is required to put these resources together, with few institutions available to do so.

National and international networks

National-level partnerships are required between:
- producers organized into viable groups
- marketers, both for the local and export markets
- processors
- suppliers of inputs
- financiers.

Whereas such institutions are already in place, they often work independently and unfortunately their aim is to get the largest possible share of the cake from the sale of the product, at the expense of the smallholder farmers. The weakest member of the chain is usually the small-scale farmer who, due to many factors, finds himself exploited by all the other members of the chain. Thus the importance of farmers' organizations is to take up the task of capacity building for the producers, with a clear focus on negotiation skills and capability enhancement.

Policies and regulations

Finally, national policies of the developing countries as well as policies from the international community should take seriously into account the participation of smallholders in the international trade:
- respective governments should put in place policies and recommendations for enhancing the long-term competitiveness of smallholder farm
- WTO agriculture issues must be addressed, such that export subsidies will be eliminated, production subsidies will be cut substantially, and barriers hindering market access will be addressed.

CONCLUSION

Smallholder farmers' participation in international trade (by themselves) is far from being regarded as development. A lot more needs to be done to enhance the necessary capacity to participate effectively in international trade. Smallholder farmers only participate in international trade by proxy as such; there is need to enhance smallholder farmers' participation in international trade through development of the relevant capacities.

CHAPTER 4

CHAINS AND NETWORKS FOR DEVELOPMENT

Articulating stakeholders in international trade

JEROEN BORDEWIJK

Senior Vice President, Supply Chain Excellence Program, Unilever N.V.
Address: PO Box 760, 3000 DK Rotterdam, The Netherlands.
E-mail: sustainable.agriculture@unilever.com

Abstract. As a major food-processing company, Unilever actively supports programmes aimed at sustainable management of its raw-material base. A large share of these materials, such as tea and palm oil, is sourced from smallholders in developing countries. Through its 'triple bottom-line approach', sourcing is linked to sustainable development. The programmes focus on the social progress of local people (people), environmental protection (planet), and economic growth in the countries that supply the materials (profit). For the sourcing of agricultural raw materials, Unilever and other food processors jointly developed a set of sustainable agricultural indicators.

The approach in reaching sustainable agricultural practices is through active stakeholder involvement. In this process, Unilever identifies innovators and agricultural organizations throughout the chain that are willing to invest in such initiatives. Knowledge gained from the initiatives is shared with these stakeholders, with the mainstream and with competitors. In discussions on how agro-food chains can work as instruments for development, Unilever feels that too often the focus is on niche markets. The real challenge is for the commodities and mainstream markets, where smallholders themselves have to organize with the support from private-sector programmes.

Keywords: sustainable development; stakeholders; supply chains; private sector

THE LINK BETWEEN BUSINESS AND SUSTAINABLE DEVELOPMENT

Unilever is a major food-processing company producing food and household products. Every day, 150 million consumers worldwide buy a Unilever product. We are present in 150 countries and have an annual turnover of 42 billion Euros. Because of this huge number of consumers world-wide, we feel that social development in the countries where we are active and where we would like to be active is an extremely important element of our business strategy. Also for our own sake: their development and their increase of income level are important to our company as well.

R. Ruben, M. Slingerland and H. Nijhoff (eds.), Agro-food Chains and Networks for
Development, 49-55.
© 2006 *Springer. Printed in the Netherlands.*

Need for sustainable economic growth

Our sustainable programmes are not about philanthropy of a company aiming at sustainable development; it is about the need to plan a business, innovate on a medium and long term, and deal with external factors that support our future. As such, environmental protection of our raw-material base is important for sustaining access to our key raw materials. Economic growth and a healthy economic development in countries where we operate create a basis for demand in these very markets; we support in developing a sustainable path to the future from which our company benefits as well.

TRIPLE BOTTOM LINE – WHAT DOES IT MEAN FOR BUSINESS

Risks and responsibilities

The future holds both risks and responsibilities for a company like Unilever. The agro-food systems are under a lot of pressure. There are economic pressures in which the chains operate, but there are also social issues. Just look at the future of the agricultural sector, both in Europe and in developing countries, and the pressures it puts on those that are active in that sector.

There are significant pressures on the natural resources too: on average 40% of agricultural land suffers from soil degradation; water supply becomes scarcer, with 70% of water supply going to agricultural use. Around 40% of the world population now faces a scarcity of water. To Unilever, water is an extremely important raw material, not only for growing agricultural products, but also for our consumers using water for our products. We are the biggest producer of tea, and our consumers cannot make a cup of tea without access to water.

There is a big challenge for scientists to make improvements that are beneficial to the crops, the environment and the people working in this environment. Take for example the genetic diversity of crops: there is a real danger that we are losing this, with the genetic base for plants becoming smaller, which in turn may have an impact on future pests that we are unable to fight.

Agriculture under pressure

For us as a company, it means that we have to look seriously at the long-term availability of some of our key raw materials, since 70% of Unilever's raw materials come out of the agro-food, fish or animal sector. We believe that we must recognize these issues. Therefore, our guiding principle is that we better become part of the solution than part of the problem.

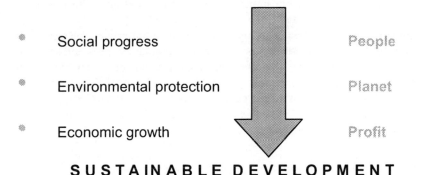

- Social progress

- Environmental protection

- Economic growth

SUSTAINABLE DEVELOPMENT

Figure 1. *'Triple bottom-line approach'*
Source: Development of international agro-food chains (Unilever 2004)

Through the 'triple bottom-line approach' (Figure 1), sourcing of required natural resources follows the Triple-P approach: work towards social progress of people, environmental protection of the planet, and economic growth or profit in the (developing) countries that supply us with the key raw materials.

Use of three key raw materials: agriculture, fish and water

Because we are the biggest users of a number of raw materials, Unilever has started specific sustainability initiatives in each of these areas. In our programmes, we focus on tackling improvements in agriculture at the location where we get the raw materials from. For the most important raw materials we started to introduce programmes in a number of countries around the world, with farmers, with local institutes and with local environmental organizations. The aim was to develop practices which will have a positive effect on soil health, on an efficient use of fertilizers and on a reduction in the use of pesticides. We also focused on improving the capacity of local people to learn and apply knowledge to the own farm, but also on becoming a better partner in the supply chain and supporting them in looking at environmental impacts, in using less water, in not polluting the water, and in ensuring that biodiversity in the areas is not negatively affected.

KEY OBJECTIVES OF SUSTAINABILITY INITIATIVES

Our consumers trust us to supply them with high-quality products that are produced in an environmentally and socially responsible way. For this reason, we act as agents to ensure that these expectations are understood along the supply chain. We must therefore align our economic goals with the social and environmental consequences of our work.

Since the mid-1990s, we have worked with stakeholders in the area of sustainability. We worked closely with environmental organizations, farmers and suppliers in our three sustainable initiatives: fish, water and agriculture. The key objectives of these initiatives are:

- Agriculture: establish sustainability indicators, appropriate measures and standards for key raw-material crops: tea, palm oil, tomatoes, peas and spinach.
- Water: support efforts to improve the conservation of clean water, and understand the water imprint of our activities. The goal is to achieve a sustainable balance between human needs and those of the ecosystem.
- Fish: encourage more sustainable fishing practices and meet our commitment to buy all our fish from sustainable sources by 2005.

Sustainable agricultural indicators

For sustainable agricultural practices, we aim at developing tailored solutions for different environments, countries and farmers. A wide range of factors can contribute to sustainable production of, for example, tea. Using a flexible framework, we strive for a balanced trade-off between agriculture, social development and conservation.

To do this we make use of ten sustainable indicators. The assessment enables tea estate companies to prioritise Sustainable Agricultural Practises (SAP) activities for their specific individual environments. These indicators are:

1. Soil fertility / Health: SAP to achieve a rich soil ecosystem.
2. Soil loss: reducing soil erosion through SAP measures.
3. Nutrients: Enhance locally produced nutrients and reduce losses using SAP.
4. Pest management: using SAP can substitute natural controls for pesticides.
5. Biodiversity: improve the diversity of biological systems through SAP.
6. Product value: SAP to improve the product value and reduce chain wastages.
7. Energy: improving the balance of energy inputs and outputs through SAP.
8. Water: using SAP to ensure water is conserved and pollutants are controlled.
9. Social / human capital: SAP to improve social (networks) and human capitals.
10. Local economy: SAP for best use of local resources (goods, labour, services).

Engaging stakeholders at every project phase

Unilever's approach in its programmes is first to engage stakeholders in every phase of these SAP projects. We look for organizations, persons and businesses that are really active and also really want to change something. We look for innovators and for organizations that are willing to take some risk. We feel that participative learning programmes are important; not telling how it should be done, but trying to jointly develop the better way, making use of innovations and use lessons-learned in pilot projects. Figure 2 shows the types of stakeholders we aim at including in our programmes.

SUPPLY CHAIN

Figure 2. Stakeholders in international food chains
Source: Development of international agro-food chains (Unilever 2004)

Agro-food chains are very complicated. In each chain; the composition of the actors and the connection between the different actors are different. This complicates the development of chains, of connecting producers in developing countries with global markets. Still, our programmes aim at increasing expertise of small farmers and help small producers to gain access to global markets.

We feel that at an early stage there is often too much emphasis on verification, on auditing, on different types of labels. We feel this is not the right approach. We should first concentrate on content, on building capacity of communities in agricultural areas to really improve but also to begin to connect to markets.

Example: tea programmes in Kenya and Tanzania

An example of this is what we do in tea in Kenya, where we try to connect to a few hundred thousand smallholders. Our aim is to provide them with expertise and know-how, try to connect them better to international markets, and let them become a partner with companies like ours, although but not exclusively ours. The programme started in 1999 on the estates of Brooke Bond Kenya, which produces 35,000 tonnes of tea on an area of over 8,000 hectares. The project teams focused their efforts on restoring and maintaining biodiversity and on optimizing renewable energy sources (hydro and fuel wood). Since an important part of the Kenyan tea is being produced by smallholders, the main challenge is to share with them the lessons learned and facilitate the adaptation process.

Another programme started in 2001 in Tanzania. Part of these tea estates need irrigation, and the programme therefore focuses on sustainable ways of irrigation as well as on the issue of soil compaction. Another important focus of the programme goes to the biodiversity conservation in the adjacent forest, which belongs to the estate but is part of a national park.

Example: palm-oil programmes in Ghana and Malaysia

Another example is related to our activities in palm oil. West Africa's lead plantation is situated in Ghana. It has a size of 4,500 ha and is managed as a smallholder operation. A set of good agricultural practice guidelines for palm oil was developed, with the challenge now being to stimulate smallholders to implement them.

In Malaysia, we have set up a round-table initiative together with the WWF, the Malaysian Palm Oil Association and a number of big plantations. We aim at sharing and building best practices, not only on how to manage plantations but also to define criteria that will be used in the process of converting rainforest into palm plantations, focusing on how to better protect the sensitive rainforest areas. Secondly, the initiative tries to define how we can create demand for sustainably grown palm oil in the chain.

Sharing knowledge with stakeholders

The knowledge we gain in these programmes is shared with stakeholders. It is also shared with competitors in the SAI Platform, a pre-competitive initiative in which food companies share sustainable agricultural practices and try to come to common approaches. It would not be good for farmers around the world when different food companies use different approaches and practices. In the SAI Platform we try to share experiences, and for a number of commodities we now have developed common programmes.

CONCLUSION

For agro-food chains to work, the power and expertise of small farmers is what we need to work on in the years to come. We need to develop know-how and provide access to expertise on the costs and the benefits of sustainable agricultural practices. Another issue can be found closer at home: sustainable initiatives need to be effectively communicated to consumers. Another challenge is how to deal with transition processes; changing the usual practice to sustainable agriculture practices involves risks. Governments and institutions can help the actors of such transition processes in minimizing or covering the involved risks.

Too often the focus of discussions on food chains as instruments for development is on niche markets. We believe that niche markets will organize themselves, and that the real challenge is in the commodities and mainstream markets. Important is to have a better forecast of supply and demand in these markets, a shift from subsidy-driven to demand-driven markets, bringing farmers closer to the market and supporting participatory learning programmes. Farmers themselves have to organize themselves; we should support them with such programmes.

For sustainable initiatives really to work, we must join forces and not have different approaches from different companies. This is a big challenge for the food industry, but also for farmers' organizations and the retail sector.

CHAPTER 5

VEGETABLES SOURCING IN AFRICA

The experience of Freshmark

JOHAN VAN DEVENTER[1]

*Managing Director
Freshmark Group*

Abstract. Freshmark is a leading South African company organizing the sourcing of fresh produce in 11 countries throughout the African continent. The company was established 15 years ago and increased its turnover from 20 million Euro in 1989 to more than 125 million Euro in 2004. Key to its success are direct eye-to-eye relations with suppliers, a fleet of refrigerated trucks and highly qualified technical and commercial staff. The main customers are the Shoprite stores that serve a market of around 30 million customers. Freshmark operations are based on the establishment of real chain partnerships, where suppliers share common goals and loyalty. Open communication as well as knowledge and understanding of each other's business reinforce truth and trust. Growers and retailers thus maintain a joint interest in promoting better product specifications through Eurepgap and HACCP norms. The core message is that information must flow from the market back to the suppliers. Local markets are highly segmented, but at the lower end of the market where people earn up to 120 Euro a month a large turnover can be made. Market access also requires joint planning towards a fine-tuned system that delivers small quantities on a daily base, with the guarantee that the produce will be collected. This guarantee enables farmers to intensify production and permits Freshmark to satisfy variety-seeking customers.
Keywords: supply chain organization; preferred suppliers; partnerships; market segmentation; South Africa

INTRODUCTION

When I was a little boy I grew up in a small railway town in South Africa, and when I became eight years old my father one day showed me on the map where the town Deventer was located. For a guy who never travelled more than 400 km in his life, Deventer in The Netherlands was really far away. And you know what is going to happen during my stay in this country: I am going to visit Deventer, 38 years later. The lesson is: never stop dreaming!

I would like to share with you the experiences of how to organize the sourcing of fresh vegetables in the African continent. We will start our journey in South Africa and then we continue into other African countries; we will soon open activities in India as well. What did we do as Freshmark group? We took the dream of Africa

R. Ruben, M. Slingerland and H. Nijhoff (eds.), Agro-food Chains and Networks for Development, 57-62.

after apartheid out of the boardroom to the African people. Today I can stand here in front of you and say: "We don't only dream about entering into Africa, we are already active in 11 African countries". And may I say right in the beginning: the objective of our Freshmark Group is to make every African country as self-sufficient as possible to get the produce from the local people to the market place.

PARTNERS IN TRADE

My main concern – and I am now talking to myself – is that we hold conferences and sometimes print very glossy brochures, but that in these African countries I sometimes miss you as a partner. I want you right next to me. Let's therefore get to reality. Let's get to where it makes a difference. And that is at ground level. There where we have an expression in South Africa: "Where the tire is the tar". That is where we can make the real difference. And therefore we need you. My invitation here is: let's take hands and let's go and change the lives of so many people through fresh produce. Because it can be done. It is a powerful tool, it can empower people, it creates jobs, we have customers, and we need the farmers that we are talking about during this conference. It can become a win-win situation.

I would like to share with you our experiences in this field. Don't re-invent the wheel. We are not the beginning and the end of the fresh-produce industry in Africa, but I might be a very good start. Where you can produce vegetables might be a very good point to start learning. Enterprises like ours can introduce the producers to the suppliers, we can take them to the farms, and this is not a dream anymore, it's no boardroom talk, it's our reality. That is my invitation and maybe next year if we have a meeting like this I would really like to say: there are the people from Angola, the farmers, there are the small-scale farmers from Malawi, to talk to them. Let them come and stand here; don't listen to me anymore.

FRESHMARK

Who is Freshmark? We are a fruit and vegetable supplier to the largest retailer in Africa: the Shoprite Group. Shoprite is the mother company, but Freshmark is an independent profit centre that is a very important part of the group. Freshmark was established 15 years ago. We grew from a turnover of 165,000,000 Rand (20,625,000 Euro) in 1989 and last year we went over one billion Rand (125 million Euro) in 2004. The scope of business are six distribution centres in South Africa, and we are present with operations in 11 other countries: distribution centres in Namibia, Zambia and Zimbabwe, and depots in Angola, Mozambique, Madagascar, Mauritius, Malawi, Tanzania, Uganda and Ghana.The procurement department in Freshmark started in 1997. Today we have 500 suppliers – farmers or growers – in South Africa and roughly 150 suppliers in the other countries of Africa. We maintain a direct eye-to-eye relation with each of them. We also have a fleet of 120 refrigerated trucks and a staff of around a thousand people. Our main customer is the Shoprite Checkers Group, operating 410 stores that need every day fresh supplies.

This represents a captured customer market of around 30 million customers per month. That is the market that we stand for.

CHAIN PARTNERSHIPS

Our operations are based on the establishment of real chain partnerships. Can suppliers become loyal supporters? Our answer is fully affirmative. And I am willing to share with you our recipe free of charge. Chain partnership is definitely feasible, if based on common goals, loyalty, truth and trust that comes from both sides.

First, it is important to know how we see each other. We sometimes consider the suppliers/farmers as coming from a different planet, since it is difficult to understand each others' motives. The typical saying *"A boer is a bok"* – and Dutch people will understand this – indicates certain stubbornness. Another expression is *"a bok is a bliksem"*, which implies that they don't really understand retail; they know nothing about marketing, market share, category management, continuous supply, etc. Farmers are usually pursuing the best prices for themselves and can tell you everything about increased input costs of labour, seed, fertilizer, transport and packaging. In this view, the trader and retailer are dictators and the suppliers can never become loyal supporters.

But can we challenge this idea that the retail is the devil, that they abuse the system and always survive, that they pay the suppliers as little as possible and ask the customer as much as possible? Is there also an alternative viewpoint possible, which looks for a partnership between growers and retailers who maintain a joint interest in promoting better product specifications through Eurepgap and HACCP norms?

We started to reconsider and break down these perceptions. Because, if these images are correct, both the farmer/grower and the wholesaler/retailer face very serious problems. We cannot be successful in today's competitive environment 'where dogs eat dogs'. Our distribution chains will not flow smoothly and will become very expensive if such distrust is maintained. Similarly, we will not be able to satisfy our customers' demands, and this is probably the most important benchmark. We should always keep in mind that Madame Customer has a choice between different retailers. In South Africa, there is fierce competition and customers have a real choice in an over-saturated retail market. Therefore, reliable supplies and constant quality are key elements in the competition.

COSTUMER ORIENTATION: KNOWING THE MARKET

The key issue to be addressed by both producers and suppliers is to maintain a focus on the same goal, which is to satisfy the demands of Madame Customer. We must know the retail market, and the core message here is that information must flow from the market back to the suppliers. In South Africa, the upper-income group that represents the wealthy people only represents about 14% of the market of any retailer. Also farmers need to know that this market segment is only 14% of the people. Although the rent value will be higher – up to 22% of the market value – in this customer category we cannot realize large turnovers. At the lower end of the market there are people who earn up to 900 Rand (119 Euro) a month, and they represent roughly 34% of the market (see Figure 1).

Knowing the market implies that suppliers should understand the structure of market demand. In the upper segments, customers in that market expect a triple P: a perfect product plus food safety, organics, ripe and ready, pre-cut for convenience, etc. But in the middle and lower segments people need basic food stuffs. They are not prepared to pay for thrills or any extras, basically because they are poor and hungry. This information regarding market composition and demand must flow on a continuous basis. Suppliers need to know in which market they operate. Customer intelligence is therefore very important; it is the beginning of everything. Retailer must also know their suppliers, because it is not sufficient just to grow the products. It is necessary to understand your customers and to follow your market. This information must flow on a continuous basis if you want to be successful.

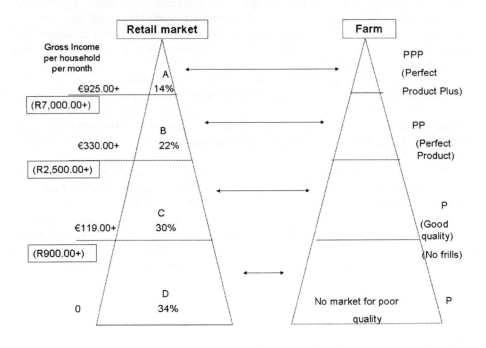

Figure 1. Market segmentation in South Africa

That upper market segment expects a perfect product, continuity of every day's supply, right packaging, right label, and so on. In addition, an A customer asks for food safety, ripe and ready, refrigerated, pre-cut, organics and this list is just getting longer by the day. On the other hand, for the D customer at the bottom end, the price is very, very important. They ask for good basic food, no poor quality, affordability, cheap prices, quick in and out, just basic delivery with no thrills and no fancies.

PLANNING TOGETHER FROM SEEDLING TO MARKETING

In other places of Africa, like in Malawi, it's a different ball game compared to South Africa. We had to cross the bridge and find each other. We first had to go to those farms in order to understand their reality – eye to eye – and to find out what is important at farm level. You cannot only talk about them and think what is the problem out there in the field; you need to go there and discuss with farmers their options and constraints.

Many farmers grow maize but have no resources to use fertilizers, and consequently the maize hardly grows and yields are extremely low. When we discussed about planting pumpkin or potatoes, they argued that they need better seeds and inputs. And they proved to be very capable farmers, proud with their first commercial harvest that provides them a rewarding income. In the board meetings we talk about business and trade, without thrills and fancies. If we want to reach the town stores, we need to plan all activities together – from seedling to marketing – to ensure that we are on the same wavelength. Planting and planning together is the key to enable smallholders to enter the market.

Planning together – row for row, basket for basket – is very well possible with small-scale farmers. We don't want to plant a hectare full of cabbage, we only want two rows. But we want these two rows every day. Therefore, our suggestion is no full specialization, but a more fine-tuned system that delivers two rows every day, with the guarantee that we will buy the produce. This guarantee enables farmers to intensify production, and it enables us to dispose of a whole variety of products for the customer.

SUPPLY-CHAIN MANAGEMENT

Today our suppliers range from big to very small. Our company maintains as common goals to cut costs, eliminate unproductive links, trading as directly as possible, improving efficiency and reducing costs by introducing returnable crates, etc. Before anything we need to know which market segments we are addressing: the A, B, C or D market? This knowledge should be with the suppliers/producers as well. When we work from the consumer back into the supply chain, this implies that we start right back with the seed selection. Supply-chain management cannot go from the farm to the customer; that is the wrong way. Management programmes need to ensure that you start from customers' demand and that everybody stays on the same track. Our keywords are: communication and knowledge of each other, understanding of each other's business, knowing your competitors. The connecting links to make this reality are truth and trust.

We can work together in a win-win relationship, letting grow the market share to satisfy the customers and of course to make money. We are not working for charity, but for business. But therefore we first need to invest at the grassroots level. We start with seeds and water supply, but this is quickly followed by a cell phone to guarantee better communication. Moreover, we want the children to go to school and receive better education.

We also learned that retail is about specifics. We cannot leave our suppliers in the dark regarding consumers' demands. We therefore give them posters with pictures of the products in real size, so that they can see what the market asks for. Anybody can take a tomato and measure it against the specifications and then can be certain that the produce will not be rejected at Freshmark. So, if you believe in good business on a continuous basis, supply-management programmes should be based on bilateral communication, knowledge and trust, and mutual understanding of each others business.

NOTES

[1] Johan van Deventer has a doctorate in Marketing and Management of the University of Pretoria and is Board member of the Shoprite Chain Group.

CHAPTER 6

BUILDING PARTNERSHIPS FOR ADDING VALUE

The role of agribusiness in developing trade

ALFONS SCHMID

*Vice-President, Food Safety and Consumer Health
Albert Heijnweg 1, 1507 EH Zaandam, The Netherlands.
E-mail: foodsafety@ahold.com*

Abstract. As a major retailer with supermarkets in many parts of the world, Royal Ahold serves millions of consumers each day. These consumers become increasingly demanding, but spend less on food than ever before. Besides safe and high-quality food, consumers increasingly want their food to be responsibly produced. Most expect the retailers to ensure this and only very few are willing to accept higher prices to do so. Retailers therefore, also in developing countries, highly depend on large volumes and low prices. Economics of scale are more important than ever, and the focus is on good and short supply chains. The focus on safe and high-quality food brings retailers to implement worldwide food safety programmes, with a network of preferred suppliers. Retailers initiate programmes where the focus is on implementing safety and good practices among their outlets (supermarkets), distribution centres and suppliers. Through independent inspections their suppliers are screened on their ability to provide safe and responsible products. Once suppliers (or organized producers) match these criteria they can benefit by becoming part of the retailer's global and regional network. In other words, by complying with the retailer's high standards at the local level, they are able to have access to global markets. To become part of a retailer's regional and global network, suppliers need to understand the competitive challenges that retailers face in serving their consumers. Retailers are looking for partners that analyse this situation and suggest solutions. Suppliers with a proactive approach can receive support from retailers by being part of their networks, and from development programmes to achieve their optimization.
Keywords: consumer demand; food quality; food safety; economies of scale; global retail networks

CHALLENGES FOR RETAILERS IN A HIGHLY COMPETITIVE ENVIRONMENT

Demanding consumers, low prices

We are the retailers, the international retailers, those that some say squeeze the prices of the products from the South, and are not willing to pay an extra for the fact that it is coming from the South, coming from a developing economy. We have all these conditions, such as EUREPGAP and the Global Food Safety Initiative, and

63

*R. Ruben, M. Slingerland and H. Nijhoff (eds.), Agro-food Chains and Networks for
Development, 63-67.*
© 2006 *Springer. Printed in the Netherlands.*

according to some we make life miserable for those who want to export their products to the areas where we operate.

This is the widespread perception among many stakeholders. Our company is Royal Ahold and we would like you to understand what our business is really about, because this is where the answer lies in being able to develop partnerships with Royal Ahold. The company has a yearly turnover of about 50-60 billion Euros, and we have around 8,000 supermarkets. These are mainly based in the USA, where we own Stop-and-Shop, Giant, Tops and Food Services, and in Europe, where we have over 600 Albert Heijn supermarkets in The Netherlands, supermarkets in Poland and the Czech Republic, and joint ventures in countries like Sweden and Portugal. We don't have one concept that applies in all countries, like some of our competitors do, but we have different formulas adapted to the local market. The most important thing is that we serve around 25 million customers every week. If we disappoint one, he or she will not come back. It is our task to understand what the customer wants, what he or she is willing to pay and what the preferences are. And that is a hell of a job.

Consumers in Europe and the US are spoiled with a great choice of products. An average supermarket in The Netherlands has 25,000-30,000 articles on the shelves; American supermarkets even have 50,000-70,000 articles on the shelves. Consumers still like high-quality and A-brands, but they are not willing to pay as much as we used to do for our food. We want to have products from all over the world: mangos all year round, strawberries all year round, so we source from Malta or Egypt, from a supplier that understands our needs.

Among consumers there is also the sense of 'guilt feeling': we want to consume and meanwhile take care of the planet. But there are only a very limited number of customers who are actually willing to pay for this feeling; most expect the supermarkets to take care of this. Supermarkets are struggling to meet with this demand without being the only ones to pay more for it. Experience has taught us that any price increase of a product in the supermarket above 5% to improve sustainability results in consumers walking away. Carrefour and Wal-Mart are dominating the retailing world, also in developing countries, and for their profits highly depend on large volumes and low prices. Nowadays, it is all about low prices. Just look at the German market: this is now completely dominated by discount supermarkets such as Lidl and Aldi. In this environment of heavy competition and low prices there is no retailer that makes a net profit of over 5 %.

In this highly competitive environment, we too need to buy large volumes to improve economies of scale, and we squeeze every single cent from the supply chain that is not really necessary. This is done by shortening the supply chains, by increasing information technology to have direct access throughout the supply chain, and by securing traceability to trace ingredients back to their roots in case something is wrong.

FOOD SAFETY AND SOURCING FROM DEVELOPING COUNTRIES

Food safety programme and standards

Food safety is an industry-wide concern. It is our top focus always to provide the safest possible products to consumers. We implemented a worldwide food safety policy in 2001, with the aim to continuously improve food safety practices at our operating units. By using the corporate 'Model Food Safety Program', local companies conduct self-evaluations and local programmes are benchmarked. Concrete action plans are then drawn up for each local company. Results are also used to facilitate the exchange of best practices through our network of suppliers. As a result, all of our local companies have stringent food safety programmes and procedures, and have ongoing improvement plans in place.

Royal Ahold is actively involved in developing the EUREPGAP standard, which is designed to ensure product safety, environmental protection, reduced use of agro-chemicals, and labour safety. The standard stands for good agricultural practices, and is an integration of the different systems being used by different European retailers. By linking these systems into one standard, certification of producers, also in developing countries, becomes simpler, more effective and cheaper. Through cooperation small producers will be able to understand that a minimum standard for us is EUREPGAP, and what this means for them. Food safety standards apply to all of our products. However, on the social, environmental and ethical issues, there is no global consensus. These issues and importance vary by country and region.

Guatemala: safe food

In Guatemala, for example, one of our companies, La Fragua, evaluated its stores and distribution centres according to our Model Food Safety Program. Based on the outcomes, a roadmap for future improvements was developed. The initiative could count on support from both the Food Safety Networking Group and the Ahold Latin-American Food Safety Committee. As a result, La Fragua improved food safety in the following areas:
- Increased staff training in safe food-handling procedures
- In-store procedures for rigorous temperature control of perishable products
- Microbiological laboratory to monitor safety and quality of perishables
- All 120 stores and distribution centres started to work towards HACCP certification.

La Fragua also launched a certification programme for suppliers of perishable produce. Through this programme, it assisted suppliers in the Good Manufacturing Standard (CFR 110 of the US FDA).

Guatemala: 'good coffee'

For our own brand of Ahold retailers in The Netherlands, Sweden, Norway and the US, the Ahold Coffee Company purchases around 15,000 tons of coffee per year. For this, we developed a structural approach for improving the social and environmental performances of its suppliers. We translated the EUREPGAP into 'good coffee' or Utz-Kapeh standards. Ahold and its coffee suppliers in Guatemala established the Utz Kapeh foundation, an independent non-profit organization that promotes the standards to local suppliers and to other retailers and roasters. In 2002, we hired a respected auditing firm to certify the coffee-growing plantations that supply the Ahold Coffee Company. This firm, working with the Utz Kapeh Code of Conduct, inspected a number of plantations on social requirements, such as minimum salaries, social security, working conditions, education for children, housing, and water and sanitation. That year, despite the very strict requirements, five plantations supplying coffee to Ahold Coffee Company were certified.

Screening of suppliers

These examples show that our companies increasingly screen suppliers on the basis of their ability to provide the safest possible products. Potential suppliers to our regional or global sourcing are asked to respond to a set of questions related to food safety on every Request for Proposal. These audits are increasingly being outsourced to independent certified inspection companies. Questions focus on assessing the supplier's methods of guaranteeing food safety and product quality, which include relevant governmental guidelines and international standards.

PARTNERSHIPS BASED ON MUTUAL UNDERSTANDING

Finding partners that provide solutions

Dialogue between farmers' organizations and retailers is important, to learn from each other what can be done and how;.to support farmers in looking at their products through the eyes of retailers, and then come up with innovative marketing ideas on meeting demands for bulk or specialty products, rather than asking us how to do it. Producers we buy from in Ghana, South Africa, Kenya and Latin America all do the marketing with us. And that is what we are looking for.

One of Royal Ahold's principles is the support of local economic development. We help to develop local businesses in ways that leverage our strengths, benefit the local economy and make sound business sense. Our companies are increasingly involved in regional and global sourcing. Through our large network of customers around the world, local suppliers can be identified as potential global or regional suppliers. To find these new international vendors, we work proactively with our operating companies. For example, local suppliers of mangos, asparagus and shrimps were able to expand their business through our regional and global sourcing network.

Our network of preferred suppliers is the Ahold Networking. It is used to enable experts around the world to access a common knowledge database, use a collective early-warning system, share knowledge, exchange best practices, and update each other on their progress in meeting individual food safety plans. We actively try to facilitate such business linkages. In Costa Rica and Guatemala, in 2002, we organized our first so-called Supplier Summit, where active local suppliers to our companies were invited to present their products to other buyers from the Ahold regional and global sourcing network.

CONCLUSION

To be part of regional and global sourcing networks, suppliers will have to understand the problems that we, as retailers, face in servicing our customers. If producers in developing countries do not understand this, if they do not understand the high level of competition among supermarkets, then they are not in the position of doing business with us. We are looking for partners who think in the same way as we do, who have analysed these issues by themselves before coming to us, and who can come up with solutions. That can be done from all over the world, but for smallholders requires a level of organization and integration. Royal Ahold supports projects in Ghana and did so in Thailand, where we try to organize activities together with organized producers. To support such activities, there is money available from development banks and governments, and there is research available from institutions and universities. Therefore, if suppliers act proactively, we are able to support them to succeed on the world market. And this, in turn, may lead to the creation of local jobs, and indirectly to the enhancement of local economic development. And at the same time, we are expanding our supplier base which enables us to purchase quality products at competitive prices for our customers.

CHAPTER 7

THE CONTRIBUTION OF FAIR TRADE TOWARDS MARKET ACCESS BY SMALLHOLDER BANANA PRODUCERS

GONZALO LA CRUZ

Representative of Solidaridad for Latin America
Address: Casilla 18-0562, Lima 18, Peru
E-mail: gonlacruz@yahoo.com

Abstract. Smallholder banana producers meet four major constraints to enter the international banana market: quality, logistics, finances and trade regulations. The quality standards are set by retailers and supermarkets and are difficult to meet by smallholders, mainly due to fungal diseases in banana. Logistics depend on availability of timely vessels over which smallholders have no control. Funding is needed for infrastructure and pre-export operations but smallholders have little access to appropriate credit schemes. The EC tariff-quota regime discriminates against non-ACP banana producers making their bananas more expensive and limiting the establishment of new import distributing companies. Fair trade (FT) is based on cost-internalization paying sustainable production practices that incorporate social rights and environmental protection. FT does not protect inefficiencies but aims to overcome the mentioned obstacles by transparent partnership between the chain partners. FT is based on higher prices paid by consumers that allow an equitable distribution of gains from trade over the chain partners. Less regulation of trade will increase the market share of FT bananas in Europe. An example of the FT organization at two levels, Agrofair and Biorganika, is given to show potential benefits and problems in developing a sustainable chain of FT and organic banana.

Keywords: quality; logistics; trade regulations; partnership; equity

THE BANANA MARKET

The export market of bananas is fully dominated by Latin America, which is responsible for 80 % of all banana exports between 1998 and 2000. In the same period the Far East, Africa and the Caribbean had shares of 13, 4 and 3 % of the export, respectively. Import markets are more diverse with the United States of America (USA) and the European Community (EC) in leading roles, responsible for 33 and 26 %, respectively, of the imports in the same period. Shares of 8, 6 and 4 %, respectively, were found for Japan, Near East, and China and Latin America (Arias et al. 2003).

R. Ruben, M. Slingerland and H. Nijhoff (eds.), Agro-food Chains and Networks for Development, 69-78.

The three largest banana trade companies are Chiquita from USA (25 %), Dole from USA (25 %) and Del Monte from United Arab Emirates/Mexico (15 %). Smaller companies are Noboa from Ecuador (11 %) and Fyffes from Ireland (8 %), while 16 % of banana trade is taken care of by a multitude of smaller companies (Chambron 2000).

Today banana trade can be divided into three major groups: conventional, 11 million tonnes; organic, 120,000 tonnes; and Fair trade (FT), 113,000 tonnes (FAO and FLO statistics).

MAIN CONSTRAINTS MET BY RURAL SMALLHOLDERS FOR ENTERING INTERNATIONAL MARKETS

The four major constraints limiting smallholders to enter international banana markets are quality, logistics, finance and trade regulations.

Quality

Quality specifications are defined by the supermarkets and retailers. Quality specifications as targets are therefore exogenous factors on which producers have little or no influence. In the banana sector, quality is defined in terms of size/diameter, age, appearance (such as colour and shine), finishing (such as free of scars and spots), flavour and smell. Quality can only be achieved through effective management of the appropriate technology throughout the entire supply chain from farming till final distribution to markets. Failing to meet the quality standards means suffering losses. Seconds cannot be exported, boxes will be dumped at port of destiny and claims will follow from ripeners and supermarkets receiving insufficient quality. One of the major constraints to meet the quality standards of the supermarkets is the lack of effective control of fungal diseases, in particular in organic banana.

Logistics

Effective logistics assure timely supply of fruit of required quality to the market. This includes all operations to achieve this objective. In the producing country, it deals with aspects from farm to port of delivery and is composed of farming, harvesting, processing, packing, palletizing, cooling, road transport and loading the vessel. Overseas logistics deal with sea transport to point of sales including unloading the vessel, transport, ripening, re-packing and delivery to shops. A critical and crucial factor is the availability of timely vessels. The smallholder producer has no control over timely sea vessels, which makes this factor a major constraint.

Finance

Funding is necessary to invest in basic facilities and infrastructure. Required funds should be available as long-term credit against reasonable interest rates. Funding is also needed as working capital for pre-export operations. It takes on average 7 weeks between the dispatch of bananas and the balance of sales from the importer. To solve this gap in time many importers make a prepayment against documents of exported fruit. Lack of funding is an important constraint.

Trade regulations

Banana imports are concentrated in two main markets with very different features:
- the USA: a free market
- the EC: a highly regulated market

Each of these markets captured around one third of the imports between 1985 and 2000 (Arias et al. 2003).

In 1993, with the establishment of the single European market, the EC put in place a regulatory regime concerning the import of bananas with the following components (European Commission 1993; 2004):
- Access to markets is regulated by a tariff-quota system.
- Quotas are allocated according to the historical volume imported by established operators
- 96.5 % of the quotas were allocated to traditional operators and 3.5 % were reserved to new ones. In 2002 new operators increased their access to quotas up to 17 %.
- Defined quota imposed a system of import licences.
- Bananas from ACP countries (Africa-Caribbean-Pacific: mainly former French and British colonies) have duty-free access to all EC member states.
- Bananas from non-traditional ACP countries (Dominican Republic and Ghana) and third countries, the so-called dollar bananas, are subject to a tariff (75 to 680 euros/tonne) according to the imported volume.
- Bananas from EU producers (mainly Canary Islands, Martinique and Guadeloupe), covered by internal aspects of the common market, enjoy an income support.

This regulated import regime has many consequences. The price of bananas for EC consumers increased by about 0.50 euro per kg since the quota reduced the volume of imported bananas. The tariff-quota system generated a market of tradable licences with the cost of a licence between 2 and 3 euros per box. The total cash value of the licences was calculated to be over US$ 1bn annually (Van de Kasteele 1998). Traditional operators benefited from this trade of licences instead of developing the banana sector. ACP countries and EC banana regions benefited from market incentives and direct subsidies.

The EC import regime favours trade of ACP bananas at the expense of the dollar bananas from non-ACP countries like Ecuador, Colombia and Costa Rica. Due to the quotas, tariffs and direct subsidies, bananas from non-ACP countries became more expensive than bananas from ACP countries.

The regulations made the entrance of new traders almost impossible. New traders need to buy licences and/or organize expensive and complex bank collaterals in order to be granted free licences by the EC. The regulated regime also increased the margin of retailers and supermarkets.

The USA and Latin-American countries, in particular Colombia, Costa Rica, Nicaragua, Venezuela, Ecuador, Mexico, Guatemala, Honduras and Panama, issued demands against the EC. Bilateral negotiations and WTO negotiations that have cost millions of euros to the EC, resulted in a modification of the tariff-quota regime so as to increase the access to import licences by dollar-banana-exporting countries. In the near future, 2006, the EC banana market will be liberalized whereby the current tariff-quota regulation will be replaced by a tariff-only system. A tariff preference will continue to be granted to ACP countries until 2008.

In summary, the major constraints are that the EC tariff-quota regime discriminates against non-ACP banana producers, making their bananas more expensive, and that the regime creates serious obstacles to new banana trade chains wanting to establish new import distributing companies.

CONTRIBUTION OF FAIR TRADE TOWARDS INCREASED ACCESS TO MARKETS

The FT system is based on the principle of cost internalization, e.g. the costs of social rights and environmental protection are included in the price paid by the consumers.

Two main separated components together contribute to FT in fruit markets:
- The rules of the game: the standards and rules for FT certification, inspection and operation. Examples are FLO, IFAT, CTM, SA 8000.
- The FT partners: the network of producers, traders, ripeners and retailers.

As an example, the FT criteria under the certification by the FT Labelling Organizations (FLO International 2004) are given. Producers are guaranteed a **minimum price** that is calculated to cover full production costs plus a reasonable margin to meet basic needs and environmental standards. The smallholder producers or workers on plantations are paid a **social premium** for further social and environmental improvements as a group. Consumers can identify a FT product (properly labelled) and pay a **higher price** for a clean and responsible product. Since 1997 FT banana knows 17 certified producers, 24 active traders and sales in 15 countries.

The FT certification, its principles and rules, can contribute towards increasing the competitiveness of the produce. FT bananas are sold to consumers at higher prices than conventional bananas. But FT can work only for a produce that has the potential to be competitive in international markets. FT does not protect inefficiencies. FT rules are a necessary but not sufficient condition. The FT scheme can set the foundation on which main constraints can be reduced so that smallholder producers can enter the market. Quality bottlenecks, weak logistics, insufficient funding and a discriminatory trade regime (Section 2) cannot be modified by FT

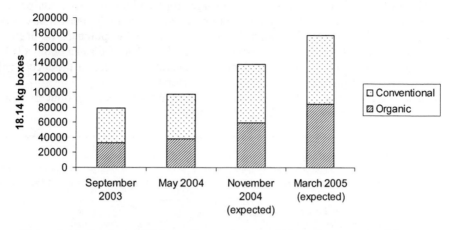

Figure 1. Market share of Fair-trade organic versus Fair-trade conventional banana (Source: FLO – number of boxes per week)

rules alone. True and transparent partnership between producers, traders and retailers is the key solution for tackling main obstacles to trade.

- Traders know about quality standards and solutions so they can provide information and technical assistance to producers.
- Retailers can modify and adapt quality requirements to existing circumstances.
- Traders (exporters and/ or importers) can help in solving logistic and sea transport problems.
- Traders can provide finance (prepayment before export and mid-term debts particularly those originated on quality claims).

FT not only corrects the price paid to producers and the price paid by consumers, transforming the chain into a feasible business, but it also allows an equitable distribution of gains from trade.

Figure 2 presents the cost structure of one box of FT conventional bananas throughout the entire chain. It is remarkable that supermarkets / retailers nearly double the cost of purchase of FT bananas to set the retail price for consumers. Therefore, about 45% of retail price is the margin of supermarkets that includes all operational and fixed costs plus their profit. On the other end of the chain, producers receive about 11% of final value paid by consumers. The US$ 1.75 premium per box (4%) is on top of all incurred costs. On average about US$ 1.5 per box could be regarded as the net extra income per box for producers due to FT.

The margin for exporters is rather small in comparison to what accrues to the other stakeholders. The net profit for exporters is somewhat between US$ 0.05 and 0.25 per box in the good cases, but there are exporters who operate close or below the break-even point.

The margin of importers is much less than the 2% of retail value indicated in Table 2. The net profit for the importer is rather variable but in any case significantly less than the net income obtained by producers due to FT and immensely less than the profit of supermarkets with FT bananas.

The FLO's national initiatives (NIs) charge importers various amounts per FT box depending on the European market, but on average US $ 0.80 per FT box. Any cost reduction during the repacking at the ripener or in the purchase of licenses mitigates this cost. Otherwise, the payment to the NIs is absorbed as an extra cost by the importer. Recently FLO is introducing a scheme of charges to producers and exporters (in addition to existing charges to importers).

In sum, FT is a system that redistributes income from traders to producers while allowing larger margins and profits to supermarkets.

Payment for fruit to producer	USD	3.45		8%
FT Premium - Investments in social and environmental programme	USD	1.75		4%
Subtotal benefit for producer	**USD**	**5.20**	**11%**	**11%**
Package, local logistics and exporter's margin	USD	1.80		4%
Subtotal FOB	*USD*	*7.00*		15%
Sea transport and insurance	USD	3.20		7%
Harbour handling	USD	0.50		1%
Overhead, financial costs and importer's margin	USD	1.00		2%
Subtotal packaging and logistic		**6.50**	**14%**	**14%**
Subtotal T1		*11.70*		26%
Import duties, licences, clearance	USD	5.00		11%
Subtotal import duties and licences	**USD**	**5.00**	**11%**	**11%**
Transport from harbour to ripener	USD	0.80		2%
Ripening	USD	1.50		3%
Packaging/pricing of clusters (USD 2.50 if applicable)	USD	2.50		6%
Distribution to DC's	USD	0.80		2%
Subtotal ripening (+packaging)		**5.60**	**12%**	**12%**
Total (is referential yellow-price for retailer T2)	**USD**	**22.30**		49%
Sales price to consumer per kg (2 Euros for FT)	USD	2.50		
Sales price per box of 18.14 kg	USD	45.35		100%
Tax VAT per box	**USD**	**2.72**	**6%**	**6%**
Retailer's margin per box	**USD**	**20.33**	**45%**	**45%**
Costs for retailer unknown (distribution, waste, store, etc.)				
Total return per box in USD at retail price		**45.35**	**100%**	**100%**
(*) Data obtained as average estimations of diverse Latin-American sources of Agrofair				
Weight per box 18.14 kg				

Figure 2. Value chain analyses for one conventional fair-trade (FT) box

Switzerland enjoys the largest market share of FT bananas in Europe. Figure 3 shows that in 2001, nearly 15% of total bananas sold in the Swiss market were FT. UK follows with 13%, The Netherlands with 12.5% and Finland with 6.5%. In recent years, the market share of FT bananas has increased in Switzerland since Coop, the supermarket chain, decided in 2004 to sell only FT bananas. It is important to note that in Switzerland the difference in price for consumers between FT and conventional bananas is very small since this country does not have the complex and regulated import regime of the EC countries and the supermarkets strongly support the FT of bananas.

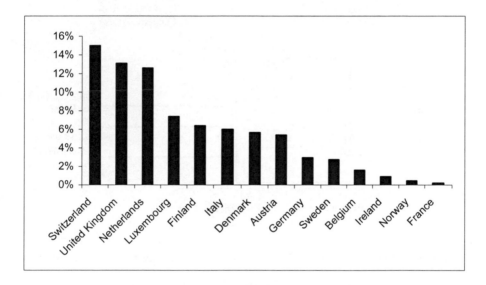

Figure 3. Market share of fairtrade bananas in Europe

FLO data show that the largest volume of FT bananas (both organic and conventional) was sold in Switzerland (data of May 2004), 39,000 boxes per week (b/w) followed by the UK with 27,080 b/w and the USA with 6,650 b/w. Other countries sell less than 4,000 b/w.

A CASE STUDY: AGROFAIR AND BIORGANIKA

In this section an example of a FT organization at two levels will be given: AgroFair Europe BV, the importer and distributor of FT fresh fruit in Europe, and Biorganika, the exporter of Peruvian bananas.

AgroFair

AgroFair Europe BV is a private trading operator that allows producers to sell their fruit directly to supermarkets. Eight producer groups from Latin America and Africa own 50% of AgroFair's shares. Foundations and NGOs (Solidaridad-Netherlands, Twin-Trading-UK and CTM-Italy) own the other 50%. In 2003 AgroFair traded 1,406,764 boxes of bananas, getting a net turnover of € 26,593.000, allowing to distribute € 200.000 as dividend for producers and NGOs.

Biorganika as exporter is a subsidiary of Agrofair in Peru. The connection between AgroFair and Biorganika is given in Figure 4.

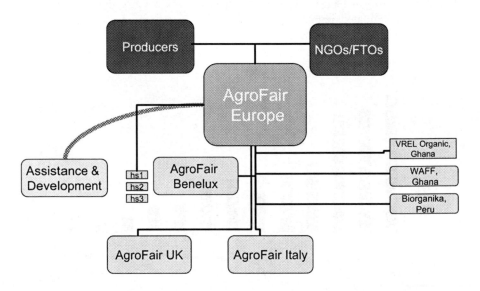

Figure 4. *Organizational structure of the holding with the position of Biorganika indicated*

Trade volumes grew rapidly from almost 60,000 boxes in 1996 to 286,000 in 1997 and 551,000 boxes in 1998. It took until 2002 before another doubling was achieved with 1,160,000 boxes. In 2003 Switzerland was responsible for 31 % of sales with Italy (28 %) in second place and UK (14 %) third. The remaining 27 % were divided over Finland (10 %), Belgium (6 %), Austria (5 %), The Netherlands and Denmark (3 % each).

Biorganika

Biorganika is a limited company acquired by AgroFair Europe BV in July 2002, owning 99 % of the shares and 1 % owned by Solidaridad. In 2003, the turnover was US $ 1.101.560 and the number of boxes traded was 125,220 equal to about 2,408 boxes per week. The company consists of 10 employees and 172 farmers (120 active). Organic banana is planted on 151 hectares, the average farm size being about 0.88 ha. BCS Oeko Garantie from Germany provided the organic certification.

Biorganika's break-even level of operation is 4,500 boxes per week providing that quality claims be kept at a minimum. During year 2004, Biorganika and the producers are working hard towards this target. It is likely that the total export for year 2004 will be about 200,000 boxes but still below the break-even point. Therefore, Biorganika will continue carrying forward losses with Agrofair's support.

What does all this mean to the smallholder producer? The minimum price for fairtrade banana was US$ 2.20 per box with an additional social premium for the Valle de Chira Association of producers of US$ 1 per box sold as fairtrade. In 2003, the Valle de Chira Association produced on average 1,035 boxes per producer. The gross annual family income originating from banana exported by Biorganika was US $ 2,590. This meant for every family an increase in income of US$ 321 per year. The social premium to the producers was US$ 80.050 for 2003 and would be twice this amount in year 2004.

MAIN CONCLUSIONS AND CHALLENGES

Looking at the constraints and the few examples of successful practices of FT, leads to the formulation of the following conclusions and challenges.

Conclusions

FT does not protect inefficiencies. Adequate quality and timely logistics are preconditions for FT: *a fair price for a fair quality product*. The FT rules of the game establish the incentives for developing trade of a produce of the highest potential. Smallholder producers can gain access to markets by means of true partnership between producers, traders and retailers (pro-active networks).

The EC import regulatory system discriminates between smallholder producers of the south by benefiting ACP countries and traditional traders. The EC import regulatory system protects inefficiencies at the expense of consumers. Less regulated markets are needed to allow further development of FT.

Challenges

Policies for FT should focus more on the compliance of minimum social and environmental standards and less on setting and manipulating prices. Policy-making on the import regime in the EC should not be based only on political criteria and relations of power among countries but policy-making should incorporate criteria of sustainable development. The EC policy should favour trade of those bananas grown under true (not artificial) sustainable and fair practices in developing countries.

Everybody can cooperate to develop FT: as consumers and/or active participants in pro-FT networks. Activities can consist of providing assistance to the development of FT chains, raising public awareness and facilitating information.

REFERENCES

Arias, P., Dankers, C., Liu, P., et al., 2003. Overview of world banana production and trade. *In:* Arias, P., Dankers, C., Liu, P., et al. eds. *The world banana economy 1985-2002.* FAO, Rome, 7-12. FAO Commodity Studies no. 1. [http://www.fao.org/documents/show_cdr.asp?url_file=/docrep/007/y5102e/y5102c00.htm]

Chambron, A.C., 2000. *Straightening the bent world of the banana.* EFTA.
European Commission, 1993. *Council Regulation (EC) No. 404/93 of 13 February 1993 on the common organization of the market in bananas.* EU, Brussels. Official Journal L 047.
European Commission, 2004. *Communication from the Commission on the modification of the European Community's import regime for bananas.* European Commission, Brussels. EC 02/06/2004 COM(2004) 399 final.
FLO International, 2004. *Fairtrade fresh fruit: guarantees a better deal for producers.* Available: [http://www.fairtrade.net/sites/products/freshfruit/why.html].
Van de Kasteele, A., 1998. *The banana chain: the macroeconomics of the banana trade.* IUF, Amsterdam. [http://bananas.xs4all.be/MacroEconomics.htm]

CHAPTER 8

THE RAPID RISE OF SUPERMARKETS AND THE USE OF PRIVATE STANDARDS IN THEIR FOOD PRODUCT PROCUREMENT SYSTEMS IN DEVELOPING COUNTRIES

THOMAS REARDON[1]

Michigan State University, Department of Agricultural Economics
East Lansing, MI 48824, USA
E-mail: reardon@msu.edu

Abstract. Supermarkets play a leading role in food supply chains in developing countries, and grades and standards are becoming key instruments for product differentiation and agri-food chain coordination. This article traces main patterns and trends in the emergence and expansion of supermarkets in developing countries, and illustrates their leading role in domestic food retail. This gives rise to a restructuring of the procurement systems of supermarkets, based on central sourcing, growing use of specialized/dedicated wholesalers and a shift towards preferred suppliers. Emerging trend also indicate a rapid rise in the implementation of private safety and quality standards in the supermarket sector for reducing the coordination costs in procurement systems. A taxonomy and illustrations of the interfaces between procurement systems and private standards is presented and implications for smallholder participation and agricultural development are discussed.

Keywords: supermarkets; procurement regimes; grades and standards; smallholders

INTRODUCTION

Standards are imposed by the various actors in the agro-food system, from 'upstream' actors such as farmers and input suppliers, to 'downstream' actors such as wholesalers, processors, retailers and food service firms. These standards can relate to: (1) quality and safety of the product itself; (2) actions to take in the production process to produce quality and safety attributes in the final product; (3) environmental and labour attributes of the production process; (4) communication such as reporting of implementation of standards. They are specified to suppliers by buyers and/or supplier organizations; if they are specified by buyers, they usually include a specification of transaction attributes such as delivery volume, timing, packaging, as well as a specification of the price and payment period.

R. Ruben, M. Slingerland and H. Nijhoff (eds.), Agro-food Chains and Networks for Development, 79-105.

Meeting the set of product and process standards and transaction attribute requirements in turn implies a specific set of practices, such as on-farm and post-harvest technologies, to be used by the producer. To gain access to a market that requires meeting a specific set of standards and other transaction attribute requirements, implies that the producer incur costs and make investments. Those expenditures are of course offset by the higher returns (relative to alternatives) of entering that market (or the supplier would not shift from his/her current market).

A key development issue thus arises. If it is necessary for a producer to shift from his/her current market (such as a stagnant rural market) to a new market (such as a dynamic urban market) with new standards in order to move out of poverty, then the producer needs to be able to meet the standards of the new market. The capacity to do so – including the broad vector of capital assets such as human, social, organization, physical and financial capital – then becomes essential to exit poverty or to manage income risk or both.

It is thus crucial to know several things: (1) is it necessary for producers to move beyond their traditional markets to access new markets in order to exit poverty and manage income risk? (2) what are the candidate 'new markets'? (3) what are the standards of the 'new markets'? (4) Who sets the standards for the new markets? (5) How are the standards implemented/enforced? (6) Can small producers meet the standards of these new (for them) markets?

This chapter will not address the first two questions, because there is a broad consensus among governments and donors in developing countries that agro-food producers need to seek export and urban markets to exit poverty and manage income risk, simply because rural incomes are not growing, or are growing much more slowly than those of consumers in the other two markets. Moreover, there is a broad consensus, reflected in a widespread adoption of agricultural diversification programmes by governments and donors, that small/medium producers need to move beyond markets for staple foods, such as bulk grains, into value-added products and non-staples such as fruits and vegetables, dairy products, meat and fish.

The chapter will, however, focus on questions 3-6 – what are the standards of the 'new markets' facing small/medium producers, who sets them and how are they implemented, and can producers meet them? I think these are among the most important development issues today.

But the focus of my approach to these questions will contrast sharply with the usual approach to these questions today. A quick scan of the literature on these issues in developing countries in the past five years focuses nearly exclusively on export markets and on public standards (standards set by governments with jurisdiction over the market in question). The literature is thus filled with debate about WTO/SPS, about non-tariff trade barriers in the form of bilateral standards, and about CODEX[2].

While I do not argue that this 'trade and public standards' focus is not an important debate, and I note that it is not useful for me to add yet another review paper on that debate, I will instead argue that that debate focuses on a relatively minor set of issues with respect to the questions above, for developing-country small/medium farmers. At least as, or even more important are private (not public) standards, set and implemented by large-scale agro-food industry firms such as supermarkets and large-scale processors (not governments or multilateral organizations) 'downstream' in domestic food markets (not export markets). The support for this point, hence the justification for my focus, is necessary.

First, the export market is on average (of course differing greatly over products) a small part of the overall agro-food market facing producers – especially small/medium producers – in developing regions. Reardon and Timmer (in press) estimate that in 2002, the share of exports in output of small/medium farmers in developing regions, is about 3% of their output, and only 5% of their marketings (of grain, fruits, vegetables, meat, fish, cotton, coffee/cocoa, sugar and oil palm). Thus trade is a very minor topic with respect to the subjects of markets and the poor in developing regions. Domestic urban markets – and who sets the terms for farmers' access to them – are a far more important subject with respect to rural development and poverty alleviation.

Second, public standards for domestic food markets are scant and scantily implemented in many developing countries. Most developing countries face public standards imposed by importing governments for export markets (such as USDA and FDA standards to export from a developing country to the US market), and governments in those countries make those local public standards imposed on exporters; the governments also usually have a plant and animal health inspection service at the border or at the ports and airports to monitor exports and imports. Some, especially the larger or more advanced developing countries, also have regulations for food safety for domestic markets. But there is plenty of emerging evidence that while these rules are 'on the books', the governments tend to have little or very little capacity to monitor domestic markets and enforce these rules. There are a few exceptions of course where the apparatus is relatively extensive, such as in China, but even there the public standards can only be enforced at key points of large-scale production such as milk-product factories.

Third, does this mean that there are no safety or quality standards applied in domestic food markets and thus the issue of 'standards in developing countries' is the most minor of development topics? Far from it. I show in this chapter that emerging very rapidly, and in many urban areas of developing countries already dominant, are 'modern' food industry firms that have the incentive – and through their procurement systems, the capacity – to implement private standards. Chief among these are supermarket chains – and the large-scale processors that supply products to meet supermarket requirements. These standards have in fact become important strategic market tools in a situation where the food market is passing from a market of commodities to a market of differentiated products heavily contested by powerful firms in consolidated food sectors.

Where the subject of the effects of private standards of modern food industry on developing-country producers has been treated in the literature, it has been nearly exclusively focused on the food industry in developed countries imposing 'export standards' on third-world exporters, such as recent work on EUREPGAP effects on produce exporters from Africa (Henson and Loader 2001), or UK supermarkets and Kenyan produce exporters (Dolan and Humphrey 2000). There has been very little research on or discussion of the effects of food-industry firms in developing countries imposing private standards on the local market, with the exception of some recent work on milk-processing private standards in developing regions[3], and the recent literature on supermarkets that is the focus of this chapter.

The gist of this paper is that the supermarket chains (and processors that meet their specifications) are the main actors in imposing private standards on producers. The thousands of traditional food industry actors downstream in the food system – including the mom and pop stores, the wetmarkets, the small-scale processors and the traditional brokers – do not have the capacity to implement standards, or to do so only minimally. But I show that supermarkets are taking over the market from the traditional players, and imposing standards that both serve to 'grow' the market and thus represent an opportunity for producers, but also imply stiff requirements for new practices by producers and thus the costs and investments noted above – and thus the possibility of exclusion of the small and poor producers from urban markets.

The chapter proceeds as follows. Section 2 lays out definitions. Section 3 traces trends in the rise of supermarkets in developing countries. Section 3 focuses on the evolution of the procurement systems of supermarkets to show that they are developing the capacity, and seeing the incentive, to implement private standards. The section starts with a discussion of organizational change in the procurement systems, and then focuses on institutional change with the development of standards and contracts, placing the latter discussion of private standards in the context of a conceptual framework. Section 4 illustrates with cases of supermarkets applying private standards. Much of the evidence is drawn from Latin America and East/Southeast Asia, the developing regions where food systems are changing most quickly and thus represent the leading edge of change to inform the debate. Section 5 discusses implications for small/medium producers and agricultural development. Section 6 concludes with policy implications.

DEFINITIONS AND ROLES OF STANDARDS

Grades and standards consist of standards ("rules of measurement established by regulation or authority") and the grades thereof ("a system of classifications based on quantifiable attributes") (Jones and Hill 1994). I use a relatively broad definition of standards, and highlight several distinctions.

First, standards can pertain to outcome or processes. 'Outcome' specifies characteristics the product is expected to have when it reaches a certain point in the agro-food chain. An example is the maximum amount of pesticide residue permitted on apples bought by a processor. Process standards pertain to any process – production of the raw product, processing into intermediate or final goods, or marketing. They specify the characteristics that the processes are expected to have, either to produce a given level of performance of the product (e.g., an organically grown apple, or meat that is safe to consume), or to create or maintain certain conditions for the environment, workers, and so on. An example of a process standard is HACCP (see Unnevehr and Jensen 1999).

Second, standards can pertain to various characteristics of a product: (1) quality (e.g., appearance, cleanliness, taste); (2) safety (e.g., pesticide or artificial-hormone residue, microbial presence); (3) 'authenticity' (guarantee of geographical origin or use of a traditional process); (4) the goodness of the production process (e.g. with respect to labour or environmental conditions). These characteristics are becoming increasingly mixed and linked, especially in private standards and procurement-management systems to implement them, which we discuss below.

Third, the standards formulating and/or implementing entity can be private or public. I do not use the term 'voluntary standard' because of the lack of operational usefulness of this term; standards imposed on suppliers by buyers are mandatory (if the supplier wants to sell to that buyer). But I use the term 'private standard' to mean a standard that is formulated and implemented by a private firm for market X. For example, this could be a safety standard for apples bought by Carrefour in Mexico for the Mexican apple market. Now, it might be that that standard is a public standard in market Y; for example, CSU Supermarkets in Costa Rica has 'CSU Standards' for fruit that are in fact a mix of private quality standards and the US-FDA fruit safety standards. But because the US government has no jurisdiction in the Costa Rican domestic market, CSU is merely using that foreign public standard as a chosen benchmark for its domestic procurement and thus I call it a private standard for the domestic market. Conversely, the Brazilian government is in the process of adopting domestic market private standards for dairy products (formulated by the large dairy-product companies) to be the public standards for dairy products. The government's standards will then be public standards even though they are based on, or benchmarked from, private standards. Finally, if a private firm merely implements a public standard (where the government has jurisdiction in the market in question), the private implementation does not make the public standard private.

Reardon et al. (1999) note that several major changes have occurred recently in the role and nature of standards, including: (1) a shift in centre of gravity from technical norms to reduce transaction costs in broad homogeneous commodity markets, to strategic instruments of product differentiation, agro-food chain coordination, market creation and share growth; (2) a concomitant shift from public toward private standards; (3) a shift from communicating experience characteristics toward reassuring consumers about credence characteristics such as food safety, worker conditions and location authenticity; (4) a concomitant shift from outcome toward process standards. These shifts are not discussed in general here, but their application in the diffusion of private standards used by supermarkets in developing regions is highlighted below.

THE RISE OF PRIVATE STANDARDS SETTERS: THE RISE OF SUPERMARKETS IN DEVELOPING COUNTRIES

A word about the focus on supermarkets

This section focuses on supermarkets as major actors in the rapidly emerging modern food industry in developing countries. This is not to suggest that supermarkets are the only formulators of private standards in these markets.

On the one hand, there is evidence that large-scale processors such as global dairy firms such as the Swiss firm Nestlé in Brazil (Reardon and Farina 2001), vegetable processors such as the Swiss firm Gerber, or cereal-processing firms like the Mexican firm Bimbo, set private standards for quality and safety of products in the developing country markets – often in advance of the specification to them of standards regarding processed products by the supermarkets, simply because they are harmonizing these standards with standards of their global operations to increase efficiency. This can lead to harmonization of private standards for processed foods over regional markets, such as in Mercosur (Farina and Reardon 2000). In the 1990s, roughly at the same time and in some cases preceding the rise of supermarkets, there was a rise of large-scale food-manufacturing firms such as those mentioned above. This often followed an initial proliferation of small and medium firms after liberalization of output markets with structural adjustment in the mid to late 1980s – and then a reconcentration of the processing sectors. The general story is told in Reardon and Timmer (in press) and there are interesting case studies such as that of Brazil (which we find to be a typical case and a front runner in trends one sees elsewhere in developing countries) in the dairy sector, told in Jank et al. (Jank et al. 1999b) and in Chile by Dirven (1999), and in Argentina by Gutman (1999), or wheat processing in Brazil, told in Farina and Furquim de Azevedo (1997).

On the other hand, large-scale processors and supermarket chains have a tendency to 'symbiosis'. Supermarket chains tend to source from large-scale processors in order to reduce transaction costs by using a few large suppliers who have adequate logistics and transportation capacity, to be assured of consistent quality and safety from companies with the capacity to monitor their quality (and enforce standards on their suppliers in turn), and to get the SKU (stock-keeping unit) range they want in 'one-stop shopping'. Examples include the Xiaobaiyang chain in Beijing shifting from 1000 to 300 processed-food suppliers as it has centralized procurement over the past two years (Hu et al. 2004), or the leading Russian chains focusing on a handful of large foreign and domestic dairy-products manufacturers for the reasons noted above (Dries and Reardon 2005). Moreover, large processors tend to want to supply to supermarket chains because the volumes are larger, their market coverage is broader (and growing rather than shrinking as with the traditional retailers), they can build product diversity and thus manage market risk through them, and supermarkets have the cold chains that the traditional retailers do not have, to handle the shift that suppliers' seek toward shorter-shelf-life products with higher margins.

The above implies that there is a 'natural' confluence of the process of private-standard formulation and implementation between the supermarket's and the large-scale processor's movement in this direction. In order then to limit the scope of this paper, we focus on the supermarket side of the equation, and make reference to this symbiosis as we proceed.

The focus here is also on supermarkets[4] because they have been largely absent from the development debate until very recently, having been traditionally viewed by development economists, policymakers, and practitioners as the retailers of rich countries or at most niche players for rich consumers in the capital cities of developing countries. But I show below that the reality has fundamentally changed, with supermarkets spreading extremely rapidly in developing countries in only the past 5-10 years (of course at different rates and depths across regions and countries) and in the process becoming the 'tail that wags the dog' of the agro-food systems in these regions. This of course does not differ from the recent experience in OECD countries, but is surprising because of the sharp difference with prevailing assumptions of policymakers in these regions, and because the process occurred so much faster than in the OECD countries, and also because, one can argue, developing-country producers are even less well-positioned than OECD farmers to deal with this shock – this change in the markets and standards they face.

In this section we describe this transformation of agro-food systems in Africa, Asia (excluding Japan), Central and Eastern Europe, and Latin America. We focus on the determinants of and patterns in the diffusion of supermarkets, and then we discuss the evolution of procurement systems of those supermarkets – the 'delivery vehicles' for private standards.

The spread of supermarkets in developing regions[5]

Determinants of diffusion

The determinants of the diffusion of supermarkets in developing regions can be conceptualized as a system of demand by consumers for supermarket services, and supply of supermarket services – hence investments by supermarket entrepreneurs. Both functions have as arguments incentives and capacity variables.

On the demand side, several forces drive the observed increase in demand for supermarket services (and are similar to those observed in Europe and the United States in the twentieth century). On the 'demand incentives' side are: (1) urbanization, with the consequent entry of women into the workforce outside the home, increased the opportunity cost of women's time and their incentive to seek shopping convenience and processed foods to save cooking time; and (2) supermarkets, often in combination with large-scale food manufacturers, reduced the prices of processed products.

On the 'demand capacity' side, several variables were key: (1) real mean per-capita income growth in many countries of the regions during the 1990s, along with the rapid rise of the middle class, increased demand for processed foods (the entry point for supermarkets as they could offer greater variety and lower cost of these products than traditional retailers due to economies of scale in procurement); and (2) rapid growth in the 1990s in ownership of refrigerators meant ability to shift from daily shopping in traditional retail shops to weekly or monthly shopping. Growing access to cars and public transport reinforced this trend.

The supply of supermarket services was driven by several forces, only a subset of which overlap with the drivers of initial supermarket diffusion in Europe and the United States. On the 'supply incentives' side: (1) as discussed below, the development of supermarkets was very slow before (roughly) the early-mid 1990s, as only domestic/local capital was involved. In the 1990s and after, foreign direct investment (FDI) was crucial to the take-off of supermarkets. The incentive to undertake FDI by European, US and Japanese chains, and chains in richer countries in the regions under study (such as chains in Hong Kong, South Africa and Costa Rica) was due to saturation and intense competition in home markets and much higher margins to be made by investing in developing markets. For example, Carrefour earned three times higher margins on average in its Argentine compared to its French operations in the 1990s (Gutman 2002). Moreover, initial competition in the receiving regions was weak, generally with little fight put up by traditional retailers and domestic-capital supermarkets, and there are distinct advantages to early entry, hence occupation of key retail locations.

On the 'supply capacity' side: (1) there was a deluge of FDI that was induced by the policy of full or partial liberalization of retail sector FDI undertaken in many countries in the three regions in the 1990s and after (e.g., partial liberalization of retail trade in China in 1992, with full liberalization of the sector scheduled for 2004, Brazil, Mexico, Argentina in 1994, various African countries via South African investment after apartheid ended in the mid 1990s, Indonesia in 1998, India in 2000). Overall FDI grew 5-10 fold over the 1990s in these regions (UNCTAD 2001); growth of FDI in food retailing mirrored that overall growth; and (2) retail procurement logistics technology and inventory management (such as efficient consumer response, ECR, an inventory management practice that minimizes inventories-on-hand, and use of internet and computers for inventory control and supplier–retailer coordination) were revolutionized in the 1990s. This was led by global chains and is diffusing now in developing regions through knowledge transfer and imitation and innovation by domestic supermarket chains.

These changes were in turn key to the ability to centralize procurement and consolidate distribution in order to "drive costs out of the system", a phrase used widely in the retail industry. Substantial savings were thus possible through efficiency gains, economies of scale, and coordination-cost reductions. China Resources Enterprise (2002), for example, notes that it is saving 40 percent in distribution costs by combining modern logistics with centralized distribution in its two large new distribution centres in southern China. These efficiency gains fuel profits for investment in new stores, and, through intense competition, reduce prices to consumers of essential food products.

Patterns of diffusion
The incentive and capacity determinants of demand for and supply of supermarket services vary markedly over the three regions, within individual countries, and within zones and between rural and urban areas at the country level. Several broad patterns are observed.

First, from earliest to latest adopter of supermarkets, the regions range from Latin America to Asia to Africa, roughly reflecting the ordering of income, urbanization, and infrastructure and policies that favour supermarket growth. The overall image is of waves of diffusion rolling along. The first wave hit major cities in the larger or richer countries of Latin America. The second wave hit in East/Southeast Asia and Central Europe; the third in small or poorer countries of Latin America and Asia including, for example, Central America and Southern then Eastern Africa. By this time, secondary cities and towns in the areas of the 'first wave' were being hit. The fourth wave, just starting now, is hitting South Asia and West Africa.

Latin America has led the way among developing regions in the growth of the supermarket sector. While a small number of supermarkets existed in most countries during and before the 1980s, they were primarily domestic-capital firms, and tended to exist in major cities and wealthier neighbourhoods. That is, they were essentially a niche retail market serving at most 10-20 percent of national food retail sales in 1990. However, by 2000, supermarkets had risen to occupy 50-60 percent of national food retail among the Latin American countries, almost approaching the 70-80 percent share for the United States or France. Latin America had thus seen in a single decade the same development of supermarkets that the United States experienced in five decades.

The supermarket share of food retail sales for the leading six Latin American countries averages 30-75 percent: Brazil has the highest share, followed by Argentina, Chile, Costa Rica, Colombia and Mexico. Those six countries account for 85 percent of the income and 75 percent of the population in Latin America. Other countries in the region have also experienced rapid growth of their supermarket sectors, but these started later and from a lower base. For example, supermarkets accounted for 15 percent of national food retail in Guatemala in 1994, and by 2002 accounted for 35 percent (Reardon and Berdegué 2002).

The development of the supermarket sector in East and Southeast Asia is generally similar to that of Latin America. The 'take-off' stage of supermarkets in Asia started, on average, some 5-7 years behind that of Latin America, but is registering even faster growth. The average processed/packaged-food retail share over several Southeast Asian countries – Indonesia, Malaysia, and Thailand – is 33 percent, but is 63 percent for East Asian countries – Republic of Korea and Taiwan. The supermarket sector in China is growing the fastest in the world; it started in 1991, and by 2003 had 55 billion dollars of sales, 30% of urban food retail, and is growing 30-40% a year (Hu et al. 2004).

Supermarket diffusion is also occurring rapidly in Central and Eastern Europe (CEE). This is occurring in three waves, with the earliest (mid 1990s) takeoff of the sector in northern CEE (Czech Republic, Hungary, Poland, Slovakia) where the share of supermarkets in food retail now stands at 40-50%. The second wave is in southern CEE (such as Croatia, Bulgaria, Romania, Slovenia) where the share is on average 25-30% but growing rapidly. The third wave is in Eastern Europe, where income and urbanization conditions were present for a take-off but policy reforms lagged, so that the share in for example Russia is still only 10% – but identified by international retailers as the number 1 retail FDI destination (Dries et al. 2004).

The most recent[6] venue for supermarket take-off, or at least pre-take-off, is in Africa, especially in Eastern and Southern Africa. South Africa is the front runner, with roughly a 55 percent share of supermarkets in overall food retail and 1700 supermarkets for 35 million persons. The great majority of that spectacular rise has come since the end of Apartheid in 1994. To put these figures in perspective, note that 1700 supermarkets is roughly equivalent to 350,000 mom and pop stores, or 'spazas', in sales. Moreover, South African chains have recently invested in 13 other African countries as well as India, Australia and the Philippines. Kenya is the other front-runner, with 300 supermarkets and a 20% share of supermarkets in urban food retail (Neven and Reardon 2004). Other African countries are starting to experience the same trends: for example, Zimbabwe and Zambia have 50-100 supermarkets each (Weatherspoon and Reardon 2003). Reardon and Timmer (in press) note that the retail transformation-lagging parts of Africa might constitute in the future a 'fourth wave'.

Second, within each of the four very broad regions there are large differences over sub-regions and countries. Usually, these can be supermarket-growth-ranked according to the variables in the supply and demand model presented above. In Latin America, for example, Brazil with a 75% share of supermarkets in food retail store sales can be contrasted with Bolivia with at most 10%; in developing Asia, Korea with 60% can be contrasted with India with 5%; and in Africa, South Africa with 55% can be contrasted with Nigeria with 5%; Hungary or Poland with shares of 40-50% can be contrasted with Russia with 10%.

Third, the take-over of food retailing in these regions has occurred much more rapidly in processed, dry and packaged foods such as noodles, milk products and grains, for which supermarkets have an advantage over mom and pop stores due to economies of scale. The supermarkets' progress in gaining control of fresh-food markets has been slower, and there is greater variation across countries because of local habits and responses by wetmarkets and local shops. Usually the first fresh-food categories for the supermarkets to gain a majority share include 'commodities' such as potatoes, and sectors experiencing consolidation in first-stage processing and production: often chicken, beef and pork, and fish.

A rough rule of thumb, applicable from Latin America, is that the share of supermarkets in fresh foods is roughly one-half of the share in packaged foods. For example, in Brazil, where the overall food retail share of supermarkets is 75 percent, the share in Sao Paulo of fresh fruits and vegetables is only 50 percent; in Argentina, the shares are 60 and 25%, respectively. This kind of rough '2 or 3 to 1' ratio appears to be typical in the regions. This difference is also common in developed countries: in France, supermarkets have 70 percent of overall food retail, but only 50 percent of fresh fruits and vegetables. The convenience and low prices of small shops and fairs, with fresh and varied produce for daily shopping, continue to be a competitive challenge to the supermarket sector, with usually steady but much slower progress for supermarkets requiring investments in procurement efficiency.

Despite the slower growth in the supermarkets' share of the domestic fresh-produce market, it is very revealing to calculate the absolute market that supermarkets now represent, even in produce, and thus how much more in other products where supermarkets have penetrated faster and deeper. For example, Reardon and Berdegue (2002) calculate that supermarkets in Latin America buy 2.5 times more fruits and vegetables from local producers than all the exports of produce from Latin America to the rest of the world.

Fourth, the supermarket sector in these regions is increasingly and overwhelmingly multi-nationalized (foreign-owned) and consolidated. The multi-nationalization of the sector is illustrated in Latin America where global multinationals constitute roughly 70-80% of the top five chains in most countries. This element of 'FDI-driven' differentiates supermarket diffusion in these regions from that in the US and Europe. The tidal wave of FDI in retail was mainly due to the global retail multinationals, Ahold, Carrefour and Wal-mart, smaller global chains such as Casino, Metro, Makro, and regional multinationals such as Dairy Farm International (Hong Kong) and Shoprite (South Africa). In some larger countries domestic chains, sometimes in joint ventures with global multinationals, have taken the fore. For example, the top chain in Brazil is Pão de Açúcar (in partnership with Casino, of France, since 1999), and the top chain in China is the giant national chain Lianhua (based in Shanghai), with some 2500 stores, in partial joint venture with Carrefour.

The rapid consolidation of the sector in those regions mirrors what is occurring in the US and Europe. For example, in Latin America the top five chains per country have 65 percent of the supermarket sector (versus 40 percent in the US and 72 percent in France). The consolidation takes place mainly via foreign acquisition of local chains (and secondarily by larger domestic chains absorbing smaller chains and independents). This is done via large amounts of FDI: for example, in the first eight months of 2002, five global retailers (British Tesco, French Carrefour and Casino, Dutch Ahold and Makro, and Belgian Food Lion) spent 6 billion bhat, or $120 million in Thailand. Wal-mart spent $660 million over 2002 in Mexico to build new stores.

These trends of multi-nationalization and consolidation fit the supply function of our supermarket diffusion model. Global and retail multinationals have access to investment funds from own liquidity and to international credit that is much cheaper than is the credit accessible by their domestic rivals. The multinationals also have access to best practices in retail and logistics, some of which they developed as proprietary innovations. Global retailers adopt retailing and procurement technology generated by their own firms or, increasingly, via joint ventures with global logistics multinationals – such as Carrefour (France) does with Penske Logistics (U.S.) in Brazil. Where domestic firms have competed, they have had to make similar investments; these firms either had to enter joint ventures with global multinationals, or had to get low cost loans from their governments (e.g. the Shanghai-based national chain), or national bank loans.

Fifth, again as predictable from the diffusion model above, the inter-spatial and inter-socioeconomic group patterns of diffusion have differed over large and small cities and towns, and over richer, middle and poor consumer segments. In general, there has been a trend from supermarkets' occupying only a small niche in capital cities serving only the rich and middle class – to spread well beyond the middle class in order to penetrate deeply into the food markets of the poor. They have also spread from big cities to intermediate towns, and in some countries, already to small towns in rural areas. About 40 percent of Chile's smaller towns now have supermarkets, as do many small-to-medium-sized towns even in low-income countries like Kenya. And supermarkets are now spreading rapidly beyond the top 60 cities of China in the coastal area and are moving to smaller cities and to the poorer and more remote northwest and southwest and interior.

DEVELOPMENT OF THE INCENTIVE AND CAPACITY TO IMPLEMENT PRIVATE STANDARDS – VIA SUPERMARKETS' TRANSFORMING PROCUREMENT SYSTEMS

We have found that supermarket chains have a dual objective – one qualitative (to increase quality and eventually safety of the product) and one quantitative (to reduce costs and increase volumes procured). Supermarket chains have a difficult time meeting those objectives by using the traditional wholesale sector to procure their products. Here is a statement from Javier Gallegos (pers. comm., 2003), the head of marketing for Hortifruti (a specialized/dedicated wholesaler for the CARHCO chain in Central America), enumerating the deficiencies of the traditional market in the face of a supermarket's needs:

> "The realities and problems of our growers and markets are as follows. The market is fragmented, unformatted, unstandardized. The growers produce low-quality products, use bad harvest techniques, there is a lack of equipment and transportation, there is deficient post-harvest control and infrastructure, there is no market information. There are high import barriers and corruption. The informal market does not have: research, statistics, market information, standardized products, quality control, technical assistance, infrastructure."

Driven to close the gap between their supplies and their needs, supermarket chains in developing regions have been shifting over the past few years away from the old procurement model based on sourcing products from the traditional wholesalers and the wholesale markets, toward the use of four key pillars of a new kind of procurement system: (1) specialized procurement agents we call 'specialized/dedicated wholesalers' and away from traditional wholesalers; (2) centralized procurement through Distribution Centres (DCs), as well as regionalization of procurement; (3) assured and consistent supply through 'preferred suppliers'; (4) high-quality and increasingly safe products through private standards imposed on suppliers.

The first three pillars (organizational change in procurement) together make possible the fourth (institutional change in procurement – that is, the rise of private standards first for quality and increasingly for safety of FFV). Below, we lay out a conceptual framework for understanding that shift, and then discuss the four pillars.

Determinants of change in supermarket procurement systems

Technology change in the procurement systems of supermarkets in developing regions is a key determinant of change in the markets facing farmers. Technology (defined broadly as physical production practices as well as management techniques) diffusion in the supermarket sector in developing countries can also be conceptualized as a system of demand and supply for new technology. Here we focus on technology for retail product-procurement systems as these choices most affect suppliers.

Demand for technology change in food-retailer procurement practices is, in general, driven by the overall competitive strategy of the supermarket chain.

However, specific choices are usually taken by procurement officers, e.g. in the produce procurement division. Hence it is crucial to understand the objective function of these officers in supermarkets in developing countries. We present a working hypothesis based on numerous interviews with these individuals.

The decisions related to purchasing products for retail shelves rests with the procurement officers in supermarket chains. Whether in the United States, Europe, Nicaragua, Chile or China, they are under several common 'pressures' from supermarket managers, operating under intense competition and low average profit margins. They are caught between the low-cost informal traditional retailers selling fresh local products on one side, and efficient global chain competitors like Walmart on the other side. The procurement officers strive to meet this pressure by reducing purchase and transaction costs and raising product quality.

Reflecting the varied demand of consumers, procurement officers seek to maintain diversity, year-round availability, and products with assured quality and safety levels.

Based on those objectives, we outline a rough model for demand (by procurement officers) and supply (by the supermarket chain to those divisions) of change in procurement systems (technology, organization, institutions). The demand function incentives and capacity variables are discussed first. Incentives include: (1) the ability of the traditional wholesale system to meet procurement-officers' objectives without the chain having to resort to costly investments in an alternative system. Usually procurement officers find this ability low, as Boselie (2002) shows in the case of Ahold for fresh produce in Thailand. Compared with the North American or the European market, produce marketing in these regions is characterized by poor institutional and public physical infrastructure support. Private infrastructure, such as packing houses, cold chains and shipping equipment among suppliers and distributors is usually inadequate. Risks and uncertainties, both in output and in suppliers' responsiveness to incentives, are high. The risks may arise due to various output and input market failures, such as inadequacies in credit, third-party certification and market information; (2) a second incentive is the need to reduce costs of procurement by saving on inputs, in this case purchased-product costs and transaction costs with suppliers; and (3) the incentive to increase procurement of products that can be sold at higher margins, hence diversify the product line into 'products' rather than mere commodities (bulk items).

Capacity to demand includes: (1) the consumer segment served by the chain. This is crucial because higher-value products cannot be marketed to poorer consumers and only cost considerations are paramount; and (2) the resources of the procurement office. These include the number of staff to manage procurement and thus ability to make organizational and institutional changes in procurement systems such as operating a large distribution centre. A variable that reflects both incentive and capacity is the size of the chain and thus product throughput in the procurement system. Usually retailers have a 'step level' or threshold throughput where they go from per-store to centralized procurement as economies of scale permit and require.

The supply of procurement technology by the chain as an overarching enterprise, to the specific product-category procurement office or offices, such as the fresh-foods categories, is an investment and is a function of several variables. The incentive variables include: (1) the importance of the product category to the chain's profits and marketing strategy. For example, we observed a small chain in an intermediate city in China that invested in building a distribution centre (DC) for processed/packaged foods but continues to buy fresh foods from the spot market (traditional wholesalers), while a national chain invested in a large DC for packaged/processed foods and has recently built a large DC for fresh foods as throughput has attained a critical mass and these products have attained a threshold importance in profits and chain marketing strategy; (2) the need for assurance of various product attributes in order to meet customers' demands (expansion of product choice, attribute consistency over transactions, year-around availability, quality and safety); and (3) the costs of the technology, such as costs of transport, construction, logistics services, etc.

The capacity variables include: (1) the size of the chain and/or access to financial capital to make the investments; and (2) the capacity of the chain to manage complex and centralized procurement systems.

The incentive and capacity determinants of demand for and supply of changes in procurement system technology vary markedly over the three regions and countries, and within countries, over chains and zones. Several broad patterns are observed in the procurement technologies that result (Reardon et al. 2003a; Berdegué et al. 2005).

First pillar of change: toward centralization and regionalization of procurement

There is a trend toward centralization of procurement (per chain). As the number of stores in a given supermarket chain grows, there is a tendency to shift from a per-store procurement system to a distribution centre serving several stores in a given zone, district, country or region (which may cover several countries). This is accompanied by fewer procurement officers and increased use of centralized warehouses. Additionally, increased levels of centralization may also occur in the procurement decision-making process and in the physical produce distribution processes. Centralization increases efficiency of procurement by reducing coordination and other transaction costs, although it may increase transport costs by extra movement of the actual products.

The top three global retailers have made or are making shifts toward more centralized procurement systems in all the regions in which they operate. Wal-Mart uses a centralized procurement system in most of its operating areas. Having centralized its procurement in France, Carrefour has been moving quickly to centralize its procurement system in other countries. For example, in 2003 and 2004 Tesco and Ahold have established large distribution centres in Poland, Hungary and the Czech Republic. In 2001 Carrefour established a distribution centre in São Paulo to serve three Brazilian states (with 50 million consumers) with 50 hypermarkets (equivalent to about 500 supermarkets) in the Southeast Region. Similarly, Carrefour is building a national distribution system with several distribution-centre nodes in China, while Ahold centralized its procurement systems in Thailand (Boselie 2002). The list goes on.

Regional chains, such as China Resources Enterprises (CRE) of Hong Kong – with Vanguard stores in southern China, are also centralizing their procurement systems. CRE is tenth in retail in China and has 17 large stores in the provinces of Shenzhen and Guangdong. In anticipation of growth following its planned $680 million investment in China over the next five years, a shift from store-by-store procurement to a centralized system of procurement covering each province is underway. Two large distribution centres were completed in 2002. The distribution centre in Shenzhen is 65,000 square meters and will be able to handle 40 department stores and 400 superstores/discount centres.

Moreover, the regional (over several countries) chains are moving toward sourcing regionally. I hypothesize that this will be, over the next decade, a factor inducing greater intra-regional trade and economic integration in regions. For example, in January 2002, a regional chain called Central American Retail Holding Company (CARHCO) was formed, composed of a Costa Rican chain (CSU Supermarkets) that had expanded into Honduras and Nicaragua, a Guatemalan chain (La Fragua) that expanded into El Salvador, and Ahold. The chain started with 253 stores in five countries and 1.3 billion dollars of sales, a large operation with about two-thirds of the supermarket sector in those countries. It started by sourcing only locally (the chain in each country mainly sourcing from local producers). However, over the past year, and with plans to increase this in the near future, the chain is starting to source regionally – say sourcing most of its dry beans from Nicaragua for the whole chain.

Second pillar of change: shift toward use of specialized wholesalers and logistics firms

There is growing use of specialized/dedicated wholesalers. They are specialized in a product category and dedicated to the supermarket sector as their main clients. The changes in supplier logistics have moved supermarket chains toward new intermediaries, side-stepping or transforming the traditional wholesale system. The supermarkets are increasingly working with specialized wholesalers, dedicated to and capable of meeting their specific needs. These specialized wholesalers cut transaction and search costs, and enforce private standards and contracts on behalf of the supermarkets. The emergence and operation of the specialized wholesalers have promoted convergence, in terms of players and product standards, between the export and the domestic food markets. Moreover, there is emerging evidence that when supermarket chains source imported produce they tend to do so mainly via specialized importers. For example, hypermarkets in China tend to work with specialized importers/wholesalers of fruit, who in turn sell nearly half of their imported products to supermarket chains. Similarly, Hortifruti functions as the buying arm of most stores of the main supermarket chain in Central America, as does Freshmark for Shoprite in Africa.

Moreover, there is a trend toward logistics improvements to accompany procurement consolidation. To defray some of the added transport costs that arise with centralization, supermarket chains have adopted (and required that suppliers adopt) best-practice logistical technology. This requires that supermarket suppliers adopt practices and make physical investments which allow almost frictionless logistical interface with the chain's warehouses. The 'Code of Good Commercial Practices' signed by supermarket chains and suppliers in Argentina illustrates the use of best-practice logistics by retail suppliers (Brom 2004). Similar trends are noted in Asia. For example, Ahold instituted a supply improvement programme for vegetable suppliers in Thailand, specifying post-harvest and production practices to assure consistent supply and improve the efficiency of their operation (Boselie 2002).

Retail chains in the three regions increasingly outsource (sometimes to a company in the same holding company as the supermarket chain) logistics and wholesale distribution function, entering joint ventures with other firms. An example is the Carrefour distribution centre in Brazil, which is the product of a joint venture of Carrefour with Cotia Trading (a major Brazilian wholesaler distributor) and Penske Logistics (a US global multinational firm). Similarly, Wu-mart of China announced in March 2002 that it will build a large distribution centre to be operated jointly with Tibbett and Britten Logistics (a British global multinational firm). Ahold's distribution centre for fruits and vegetables in Thailand is operated in partnership with TNT Logistics of The Netherlands (Boselie 2002).

Third pillar: toward preferred-supplier systems

Many supermarket chains are undertaking institutional innovation by establishing contracts with their suppliers – in particular via their dedicated, specialized wholesalers' managing a preferred-supplier system for them. This trend is similar to that in agro-processing during the past decade (Schejtman 1998). The contract is established when the retailer (via their wholesaler or directly) 'lists' a supplier. That listing is an informal (usually) but effective contract[7], in which delisting carries some cost, tangible or intangible. We have observed such contracts in all the regions under study. Contracts serve as incentives to the suppliers to stay with the buyer and over time make investments in assets (such as learning and equipment) specific to the retailer specifications regarding the products. The retailers are assured of on-time delivery and the delivery of products with desired quality attributes.

These contracts sometimes include direct or indirect assistance for farmers to make investments in human capital, management, input quality and basic equipment. Evidence is emerging that for many small farms these assistance programmes are the only source of such much valued inputs and assistance – in particular where public systems have been dismantled or coverage is inadequate. In some cases, the assistance is indirect – such as the case of Metro supermarket chain (a German chain) in Croatia intervening with the bank (noting that the suppliers would have contracts) to provide a 'collateral substitute' so would-be strawberry suppliers could make needed greenhouse investments (Reardon et al. 2003b). This constitutes resolution by retailers or their wholesaler agents of idiosyncratic factor market failures facing small producers – such as credit, information, technical assistance, and so on. There is evidence of this in the processing sector also, for example in the CEE (Gow and

Swinnen 2001; Dries and Swinnen 2004). Some cases of this are remarkable in their extent and nature. Codron et al. (2004) note a case of a Turkish retailer MIGROS which contracts with a whole village nearby its Antalya market to grow 1000 tons of tomatoes during the summer. Hu et al. (2004) describe the case of Xincheng Foods in Shanghai, acting as a specialized wholesaler for the top two chains in China. Xincheng long-term leases (from townships) 1000 hectares of prime vegetable land, hires migrant labour, installs greenhouses and uses tractors and drip irrigation (thus changing production technology), and produces in-house large quantities of high-quality vegetables for the supermarket chains and export. It also has contracts with 4500 small farmers to add to its own production. This kind of operation can be described as a major 'agent of change' in the Chinese agro-food economy.

While the contracting is quite recent for produce, it has been a practice for a half decade or more among chains sourcing from processed-product suppliers. Manufacturers of private-label processed fruit and vegetable and meat and cereals products typically operate under formal contract with the supermarkets. Supermarket chains have contracts with processing firms, which in turn may sign contracts with producers. For example, the processing firm IANSAFRUT supplies processed vegetables to supermarkets in Chile under such an arrangement (Milicevic et al. 1998). Similarly, processed fruits and vegetables are sold under the label SABEMAS for the supermarket CSU in Costa Rica, and various firms produce under contract the products for the private label. As retail sales of private label products continue to grow, such contract arrangements are expected to increase in Latin America and Asia.

Fourth pillar: the rise of private standards

While food retailing in these regions previously operated in the informal market, with little use of certifications and standards, the emerging trend indicates a rapid rise in the implementation of private standards in the supermarket sector (and other modern food industry sectors such as medium/large-scale food manufactures and food service chains). The rise of private standards for quality and safety of food products, and the increasing importance of the enforcement of otherwise-virtually-not-enforced public standards, is a crucial aspect of the imposition of product requirements in the procurement systems. In general, these standards function as instruments of coordination of supply chains by standardizing product requirements over suppliers, who may cover many regions or countries. Standards specify and harmonize the product and delivery attributes, thereby enhancing efficiency and lowering transaction costs. In turn, the implementation of these standards depends crucially on the establishment of the new procurement-system organization noted in the three pillars above.

Below we lay out a conceptual framework for the diffusion of private standards among supermarkets, and then provide a taxonomy and illustrations of their use.

Conceptual framework for the diffusion of private standards

The usual technology-adoption model has adoption as a function of a vector of incentive variables (relative output and input prices and risk) and a vector of capacity variables, reflecting the would-be adopter's capacity to respond to incentives (capital assets such as human, organization, physical, social and financial capital), and various 'shifters'. This general adoption framework can be applied to 'institutional adoption' such as the adoption of private standards by supermarket chains' procurement arms or agents in developing regions.

The incentives include the following.

First, the chain has an incentive to implement private standards where there are missing or inadequate public standards, so that private standards are a substitute for the missing institution. As the large chains (and processing firms) competed in national and regional markets and attempted to differentiate their products to protect and gain market share, they found that the public standards needed for that differentiation did not exist (common in developing regions, see Stephenson 1997), or relatively undifferentiated public standards existed, inherited from the protected, homogeneous commodity markets that were common before market liberalization and structural adjustment. The latter were inadequate either to meet consumer demand for product differentiation and quality differences, or to reward producers for their investments in quality and safety (Reardon et al. 1999; Reardon and Farina 2001). As noted above, governments in these regions tend to have the incentive and capacity to implement public standards mainly for the export-market interface, and much less so for domestic markets. Moreover, public standards tend to be applied where they are 'public goods' such as for plant and animal health. At the opposite extreme are quality standards that are typically private goods, differentiating products, and are the first and foremost domain of private standards.

Between the two are food safety standards. In principal, these should be considered public goods and set and enforced by governments. The issue here is not conceptual but rather practical – governments might occasionally establish regulations but usually do not have the capacity to monitor and enforce them (for the case of Guatemala, see Flores-Navas 2004). Yet supermarket chains have incentives to set private safety standards, at least for 'at risk' products such as leafy greens, berries and other products where pesticide residuals and bacteria can produce short-medium-run health problems among their clientele. In some countries there are liability laws that make this a legal issue. Yet even where there are not laws, there are two other reasons to have such standards. On the one hand, as noted above, most of the chains are global or regional, and a health crisis caused by an unsafe product in one country can hurt sales and stock prices in the region or globally. On the other hand, safety standards – and the belief on the part of the consumer that chains are able to actually monitor and enforce them – gives a big advantage to supermarkets over traditional retailers, and thus is a major competitive instrument.

Of course, where there are public standards for safety, private standards can meet or exceed the stringency of public standards thus affording 'domain defence', limiting exposure to penalties from public regulations (Caswell and Johnson 1991). Communicating to the urban or developed country consumer that the private standards exceed the stringency and enforcement of public G&S encourages consumers to buy products from countries that they may see otherwise as having lax quality and safety regulations.

Second, private standards are used to increase profits through facilitating product differentiation – and thus provide incentives to suppliers to make asset-specific investments, and to consumers to satisfy their desire for product diversity by shopping at the chain. Supermarkets (as well as large-scale processors and fast-food chains[8]) use private standards to differentiate their product lines (adding SKUs and thus product diversity) and differentiate their products from each other and from traditional actors. Private standards make product differentiation easier and more flexible, allowing companies to take advantage of new market opportunities ('domain offense', Caswell and Johnson 1991). Consistent implementation of private G&S, plus certification, labelling and branding systems that link high quality and safety standards to the product and the company in the consumer's mind, produces reputation and competitive advantage. One sees this in the application of the Carrefour Quality Certification programme and labels for meat and produce in Mexico, China, Brazil and elsewhere.

Third, chains use private standards to reduce cost and risk in their supply chains. The main cost reduction comes from using process standards to coordinate chains. Farina (2002) and Gutman (2002) illustrate these cost savings in the case of supermarkets and dairy products in Brazil and Argentina. Chains complement private standards with other elements of a "metasystem of quality control" (Caswell et al. 1998), adding elements such as branding to the system governance structure. Building trust and reputation around the visible symbol of a brand name and label makes standards systems credible to consumers (Northen and Henson 1999). To build consumer confidence (and thus build market volume and reduce market risk) by consistency in standards implementation, tight vertical coordination is needed, especially for process standards – hence the use of the organizational structure of procurement, plus contracts, noted above.

An important element of this is the reduction of coordination costs in procurement systems that become progressively broader in geographic scope, as the discussion of the first pillar above establishes as a trend. Regional and global chains want to cut costs by standardizing over countries and suppliers as this occurs – which induces a convergence with the standards of the toughest market in the set, including with European or US standards. One sees this in Wal-mart between Mexico and the US, one sees this in the Quality Assurance Certification used by Carrefour over its global operations that include developing countries, one sees this in the regional chains such as CARHCO discussed above. In some cases this has meant that global chains actually apply public standards from their developed-country markets as private standards to suppliers to their local developing-country markets, such as the use of FDA standards for some products by US chains. The chains might also use private standards from the developed country portions of their markets, such as European chains using EUREPGAP standards for some produce and meat items applied to suppliers in developing-country markets.

The capacity variables involved in the diffusion of private standards are as follows.

First, the chains, or their specialized/dedicated wholesalers, must have the requisite degree of buying power to impose private standards on suppliers – either because the chain has some oligopsonistic power, or because it offers higher producer prices, or it offers other assistance to producers. The size of the front-runner chains (the same ones that are the main implementers of private standards) relative to the urban market certainly gives them the buying power (for example, Carrefour has about 25% of all food retail in Argentina, Wal-mart has 20% of all food retail in Mexico).

Large chain size is necessary but not sufficient – as chains need the procurement organization changes noted above, in particular distribution centres that allow the product procurement to be centralized allowing efficient standards monitoring, and implicit contracts (via the preferred-supplier systems) that allow traceability and a delivery vehicle for the standards.

Sometimes chains also offer prices higher than the wholesale-market prices to producers who meet their standards; little systematic information exists about this point, but in general we have found that the premium is around 10-15%, just enough to meet additional costs implied by meeting the standards. But sometimes no price premium is offered: what then is the incentive for the producer to meet the (usually more stringent) private standards? The answer is related to the discussion of the preferred-supplier systems above: chains (or their specialized/dedicated wholesalers) sometimes offer technical assistance, input credit or collateral substitutes in the form of a contract, and transport to their suppliers. (An example is Hortifruti's technical assistance and credit to vegetable suppliers in Costa Rica.) The technical assistance and credit resolve idiosyncratic factor market failures that often plague producers after public systems for these items were dismantled during the structural adjustment period – and one can hypothesize that public systems were never nor are now adequate to meet the kinds of upgrading needs that face suppliers to supermarkets.

Second, all of the above is necessary but not sufficient to implement private standards – the final ingredient is the capacity of producers to meet the standards. A poignant illustration of this was the limitation felt by the La Fragua chain in Guatemala to implement broadly its new 'Paiz Seal' quality and safety certification system in the past two years. They found the following: (1) for key bulk items such as Roma tomatoes, there were simply not enough producers with the capacity to supply over the full year or sufficient volume to meet the chain's needs, and so the chain has to rely on traditional wholesalers to bulk the product from many small producers – obviating traceability and imposition of safety standards and quality consistency; (2) for key 'at risk' items such as leafy greens and berries, the chain has been forced to take a gradual approach of approving suppliers, at a rate much slower than it wanted, simply because few producers can make the needed investments, and those producers have export-market alternatives. Because of these limitations on finding enough suppliers that can meet the private standards, some chains take a position in between no application of standards and full, rigorous application. For example, CSU Supermarkets/Hortifruti in Costa Rica monitors standards compliance, but then is loathe to 'delist' suppliers who violate standards, even safety standards. Instead, when a problem is identified, they increase technical assistance combined with warnings, with some eventual delisting (hence, the combination of a carrot and stick approach, but not too stern so as to find themselves with inadequate supply) (Berdegué et al. 2005).

Taxonomy and illustrations of interfaces between procurement systems and private standards
In this section I draw from a Central American illustration in Berdegué et al. (2005). The degree to which this overall model of procurement systems is described by the 'four pillars' above varies across the region, and across chains. The sequence here is from the 'traditional procurement system' of Central American supermarkets (decentralized, relying on traditional wholesalers), to modern systems with an emphasis on the four pillars discussed above.

Type 1: Total reliance on traditional wholesalers delivering to individual stores. A few relatively small chains and all the independent supermarkets, such as Unisuper in Guatemala (12 medium-sized and 12 relatively small supermarkets) or La Colonia in Nicaragua (7 stores), continue to rely on the traditional system in which traditional wholesalers deliver produce to each individual store and only minimal quality standards are applied (requesting sorting from the wholesalers). In these chains, quality standards are low (basically relying on what is available that day in the wholesale market) and their control is based on rejecting high proportions of wasted produce after it can no longer be sold.

Type 2: Outsourced and decentralized procurement system. This is a system utilized by small-medium chains, such as Megasuper in Costa Rica (with 15% of the supermarket market) or PriceSmart in Costa Rica, Honduras and El Salvador (with a few stores in each country). These chains lack the critical mass in terms of produce sales, to justify a centralized operation. Instead, they rely on one or two specialized wholesalers, who in turn source mostly from the central wholesale markets and, in some products, from individual growers. Quality standards are higher in this system than in the previous one, both because the chains are larger and, in some cases, are focused on a middle-high- to high-income clientele (e.g., that of PriceSmart), and because the specialized wholesalers are also stronger and fully formal firms, as compared to the traditional wholesalers that are common in type-1 procurement systems. Yet, quality standards in this type 2 are still strictly limited to cosmetic and flavour characteristics, as much of the supply is coming from the central markets, and it thus becomes impossible to control for variables other than those that can be appreciated rapidly by simply looking at the product.

Type 3: Decentralized mixed procurement system. This type of arrangement can be found in chains which are about to make the switch to a centralized procurement system. An example is that of SuperSelectos in El Salvador (which is tied for first place with La Fragua, with about 55 supermarkets and a chain of small-format stores). The chain still is largely reliant on one or two specialized wholesalers. From one wholesale company, Gladys de Alvarado, it gets 70% of its regional produce, nearly all from Guatemala; Gladys de Alvarado has, in turn, a system of preferred suppliers in Guatemala and also buys from the wholesale market and from other specialized wholesalers there. However, SuperSelectos itself still has a significant complement of direct sourcing from individual growers and from preferred wholesalers/suppliers in the central wholesale markets. Relying on more than one supplier gives more leverage to the chain to demand higher quality and lower price from the main specialized wholesaler. Thus, quality standards tend to be higher than in the more standard type-2 system and the type-1 system, but again limited to those characteristics that can be evaluated rapidly and simply by expert observation.

Type 4: Centralized passive procurement system. This arrangement allows the chain to define and enforce much stricter quality as well as begin, in a limited subset of producer-suppliers and products, to implement safety standards, including, for example, standards on pesticide residues or presence of pathogens such as *E. coli*. The best example in the region is that of La Fragua in Guatemala.

La Fragua, with its various formats (such as Supermercados Paiz and HiperPaiz), has 65% of the supermarket sector in Guatemala. La Fragua has also moved in the past five years to centralize its FFV procurement through its subsidiary Disfruve. In 1999, only 20% of its procurement was 'centralized' (procured and then distributed to the stores through the small warehouse at Disfruve) – and by end 2004, 98% of its procurement is centralized (through its large, modern DC built in 2002). In 1999, about 25% of its FFV came from producer-suppliers (as opposed to wholesaler-suppliers delivering from rural areas or from the wholesale market) – and by end 2004 more than 40% comes from producer-suppliers. During the five years, the volume moved by Disfruve quintupled to keep pace with the rapidly growing chain. The combination of centralization and progressive shift toward use of producer-suppliers (sourcing directly) is providing Disfruve with a growing capacity to enforce more stringent quality standards at lower monitoring cost. The standard has been formalized in writing for each product, and a well-trained group of employees receives and inspects each shipment. Those with the highest rates of compliance get rewarded with orders for increased volumes of FFV during the next weeks, and the opposite happens to those suppliers who perform less well.

We call this a passive procurement system because from the point of view of La Fragua, it is up to the supplier to meet its rules and to find the best way to do so. The chain simply sets out clear rules and a monitoring, enforcement and incentive system.

Here is the point in this continuum of development of procurement organization and institutions where produce safety standards make their first appearance. La Fragua has seen the incentive to move one step further and establish in June 2003 a formal quality and safety seal, the 'Paiz Seal' (after its main chain, Paiz). This retailer produce-safety seal is conferred on producers who agree to sell the products with the seal only to La Fragua, and who pass the test of the third-party certification scheme, PIPAA. La Fragua wants to move the above safety/quality standard/seal from voluntary to mandatory over the next year or two. At present, however, it plans on continuing the 'passive' system where it is the choice, responsibility – and burden – of the supplier to meet the production and post-harvest level requirements of this certification. There is no premium planned, only preference in sourcing and eventually access to sales.

Another transition point is occurring in this system: starting in mid 2003, La Fragua started (albeit with a small share, about 10%, of its preferred producer-suppliers) to shift toward a combination 'passive/active' system by hiring an agronomist to train producers in Good Agricultural Practices toward obtaining and maintaining the certification; by March 2004, 25 medium-sized growers had obtained the certification, in particular for 'high risk' products such as salad tomatoes, bell peppers, endives, lettuce, pineapples, carrots and strawberries.

Type 5: Centralized proactive procurement system. The major difference between this system and the previous one is that in this case the supermarket chain establishes a technical assistance and training programme to help its suppliers in making the transition to higher quality and safety standards. The only example in the region is that of CSU supermarkets. CSU has 80% of the supermarket sector in Costa Rica. Since 1972 CSU has relied on a specialized, dedicated wholesaler, Hortifruti, for its FFV procurement. Hortifruti is a company in the same holding company as CSU.

Until about eight years ago, Hortifruti relied mainly on the traditional wholesale market, buying in bulk, delivering lots to its DC, then breaking down the lots and sending small lots around to the CSU stores. As CSU grew into a chain of 97 stores in Costa Rica, the need to procure large volumes and standardize quality became crucial. Over the past 3-4 years Hortifruti moved nearly fully away from reliance on the traditional wholesale market (today it only buys 15% of its produce from the wholesale market and only 10% from imports via a specialized fruit importer).

But Hortifruti went a step further. Under the impetus of closing the price gap with wetmarkets that was impeding their penetration of the FFV market in Costa Rica, and increasing the quality gap, Hortifruti combined the above shift, with the establishment of a network of approximately 200 preferred FFV suppliers. Fifty of these are mainly fresh-processors (such as of fresh cuts) and grower/packers that aggregate product from other suppliers. The rest are individual growers or grower/packers. Each supplier must clean, crate or pack in final usable trays the product, and deliver to the Hortifruti DC. The attraction for the growers is the promise of stable access to an attractive and growing market, at prices that are close to but usually a bit above the wholesale market, plus technical assistance, and for the small farmers, input credit. In May 2003 Hortifruti conferred on a tenth of their producers, mainly medium farmers producing leafy greens, the Hortifruti Quality Seal, which essentially combines the public Sello Azul (for low pesticide use) with Codex standards for *E. coli* plus Hortifruti private quality standards.

IMPLICATIONS FOR PRODUCERS AND AGRICULTURAL DEVELOPMENT: OPPORTUNITIES AND CHALLENGES FOR PRODUCERS FROM SUPERMARKETS' PRIVATE STANDARDS

Meeting private standards can present clear opportunities for producers. Adopting standards can open the door to suppliers of selling through supermarket chains that are 'growing' the market in terms of volume, value-added and diversity. A supplier can move from being a local supplier to a national, regional or global supplier. Moreover, private process standards can increase efficiency of firm operations and raise profitability (Mazzocco 1996). The market scope could also increase, compensating for per-unit profit decreases arising from costs incurred to meet the standards.

However, meeting new, more stringent private standards (compared to the traditional system) implies changes in production practices and investments, such as reducing pesticide use and increasing IPM use on farms, or investing in 'electric eyes' in packing sheds and cooling tanks in dairies. Some of these investments are quite costly, and are simply unaffordable by many small firms and farms. It is thus not surprising that the evidence is mounting that the changes in standards, and the implied investments, have driven many small firms and farms out of business in developing countries over the past 5-10 years, and accelerated industry concentration.

The supermarket chains, locked in struggle with other chains in a highly competitive industry with low margins, seek constantly to lower product and transaction costs and risk – and all that points toward selecting only the most capable farmers, and in many developing countries that means mainly the medium and large farmers. Moreover, as supermarkets compete with each other and with the informal sector, they will not allow consumer prices to increase in order to 'pay for' the farm-level investments needed to meet quality and safety requirements. Who will pay for water-safe wells? Latrines and hand-washing facilities in the fields? Record-keeping systems? Clean and proper packing houses with cement floors? The supplier does and will bear the financial burden. As small farmers lack access to credit and large fix costs are a burden for a small operation, this will be a huge challenge for small operators.

It is thus inevitable that standards demanded by consumers are increasingly a major driver of concentration in the farm sector in developing regions. As supermarkets' direct share in the FFV market grows, and as their influence is increasingly felt on the practices of informal markets through competition for the most profitable clients (the middle- and high-income segments) and consumer expectations, the effect of rising standards will spread over the farm sector. While it is very probable that this means that consumers will consume fewer pesticides and harmful microbes, and have better-quality food products, it also means that development programmes, in the context of weak public support systems for agriculture, will have a challenge and a mandate to assist small farmers to make the transition.

NOTES

[1] Thomas Reardon is Professor of Agricultural Economics at Michigan State University. This chapter draws on a series of collaborative research papers on this subject as well as a draft paper prepared for the OECD.

[2] Several strands stand out: (1) use of public G&S as non-tariff trade barriers against tropical products (e.g., see ECLAC 1998), for Latin America, and Henson and Loader (2001), in general); (2) trends, and in particular, difficulties in harmonization of public G&S in developing regions (e.g., see Stephenson 1997); (3) an incipient literature on the rise of process G&S and their costs of implementation for poor countries and small firms (e.g., see Diaz (1999) in general, and Deodhar and Dave (1999), for India).

[3] See for example Farina et al. (2005); Farina (2002); Gutman (2002); Dirven (1999); Jank et al. (1999b); Dries and Swinnen (2004); Farina et al. (2005).

[4] This is a term we use as shorthand for large-format modern retail stores, such as supermarkets, hypermarkets and discount stores. Our discussion focuses on large format because convenience stores tend to have only a tiny share (3-5%) of modern retail-sector sales.

[5] This section and the next draw on several publications – in particular on Reardon and Timmer (in press) and Reardon et al. (2003a) for overall trends, and also from other papers such as for Latin America, Reardon and Berdegue (2002), Balsevich et al. (2003) and Berdegue et al. (2005), for Central and Eastern Europe, Dries et al. (2004), for China, Hu et al. (2004), and for Africa, Weatherspoon and Reardon (2003) and Neven and Reardon (2004).

[6] South Asia is poised at the edge of a take-off, with the share of supermarkets in India at 5%, but identified as number 2 in the top 10 destinations for retail FDI today (Burt 2004).

[7] 'Contracts' is used in the broad sense of Hueth et al. (1999), which includes informal and implicit relationships.

[8] It has been common for processing firms to create private standards to replace or sidestep public standards and grading systems. Zylberstajn and Neves (1997) and Farina and Furquim de Azevedo (1997) illustrate this for coffee and wheat products in Brazil, Jank et al. (1999a) for dairy products in Brazil, and Farina (2002) for the Nestlé Quality Assurance certification programme for coconut products in Brazil.

REFERENCES

Balsevich, F., Berdegué, J.A., Flores, L., et al., 2003. Supermarkets and produce quality and safety standards in Latin America. *American Journal of Agricultural Economics,* 85 (5), 1147-1154.

Berdegué, J.A., Balsevich, F., Flores, L., et al., 2005. Central American supermarkets' private standards of quality and safety in procurement of fresh fruits and vegetables. *Food Policy,* 30 (3), 254-269.

Boselie, D., 2002. *Business case description: TOPS supply chain in Thailand.* Agrichain Competence Center/KLICT, Den Bosch.

Brom, F., 2004. *Design of a code of best commercial practices to enhance relations among supermarkets and food suppliers in Argentina: presentation at the International conference Supermarkets and agricultural development in China: opportunities and challenges, Shanghai, May 24-25 2004.* [http://cati.csufresno.edu/cab/rese/ShanghaiConf/Fernando%20Brom/Fernando%20Brom-paper.doc]

Burt, T., 2004. Global retailers expand markets. *Financial Times* (June 22), 15.

Caswell, J.A., Bredahl, M.E. and Hooker, N.H., 1998. How quality management metasystems are affecting the food industry. *Review of Agricultural Economics,* 20 (2), 547-557.

Caswell, J.A. and Johnson, G.V., 1991. Firm strategic response to food safety and nutrition regulation. *In:* Caswell, J.A. ed. *Economics of food safety.* Elsevier, New York, 273-297.

China Resources Enterprise, 2002. *Retailing strategies and execution plan.* China Resources Enterprise, Hong Kong. [http://www.cre.com.hk/presentation/2002GlobalInvestors.pdf]

Codron, J-M., Bouhsina, Z., Fort, F., et al., 2004. Supermarkets in low-income mediterranean countries: impacts on horticulture systems. *Development Policy Review,* 22 (5), 587-602.

Deodhar, S.Y. and Dave, H., 1999. *Securing HACCP in agriculture to build brand image: case studies from India: paper presented at the International workshop on markets, rights, and equity: rethinking food and agricultural G&S in a shrinking world, Michigan State University, 1-3 November.*

Diaz, A., 1999. *La calidad en el comercio internacional de alimentos.* PROMPEX, Lima.

Dirven, M., 1999. Dairy clusters in Latin America in the context of globalization. *International Food and Agribusiness Management Review,* 2 (3/4), 301-313. [http://www.ifama.org/nonmember/OpenIFAMR/Articles/v2i3-4/301-313.pdf]

Dolan, C. and Humphrey, J., 2000. Governance and trade in fresh vegetables: the impact of UK supermarkets on the African horticulture industry. *Journal of Development Studies,* 37 (2), 147-176.

Dries, L. and Reardon, T., 2005. *Central and Eastern Europe: impact of food retail investments on the food chain.* FAO Investment Center, London. EBRD Cooperation Program Report Series no. 6.

Dries, L., Reardon, T. and Swinnen, J., 2004. The rapid rise of supermarkets in Central and Eastern Europe: implications for the agrifood sector and rural development. *Development Policy Review,* 22 (5), 525-556.

Dries, L. and Swinnen, J.F.M., 2004. Foreign direct investment, vertical integration, and local suppliers: evidence from the Polish dairy sector. *World Development,* 32 (9), 1525-1544.

ECLAC, 1998. *Non-tariff trade barriers to exports from the LAC region.* United Nations Economic Commission for Latin America and the Caribbean, Santiago de Chile.

Farina, E.M.M.Q., 2002. Consolidation, multinationalisation and competition in Brazil: impacts on horticulture and dairy product systems. *Development Policy Review,* 20 (4), 441-457.

Farina, E.M.M.Q. and Furquim de Azevedo, P., 1997. Moinho pacifico: ajustamentos e desafios do Livre-Mercado. *In:* Farina, E.M.M.Q. ed. *Estudos de caso em agribusiness.* Biblioteca Pioneira de Administracao e Negocios, Sao Paulo, 25-41.

Farina, E.M.M.Q., Gutman, G.E., Lavarello, P.J., et al., 2005. Private and public milk standards in Argentina and Brazil. *Food Policy,* 30 (3), 302-315.

Farina, E.M.M.Q. and Reardon, T., 2000. Agrifood grades and standards in the extended Mercosur: their role in the changing agrifood system. *American Journal of Agricultural Economics,* 82 (5), 1170-1176.

Flores-Navas, L.G., 2004. *Small lettuce farmers' access to dynamic markets in Guatemala.* Michigan State University. Masters thesis, Michigan State University

Gow, H.R. and Swinnen, J.F.M., 2001. Private enforcement capital and contract enforcement in transition economies. *American Journal of Agricultural Economics,* 83 (3), 686-690.

Gutman, G.E., 1999. Desregulacion, apertura comercial y reestructuracion industrial: la industria lactea en Argentina en la decada de los noventa. *In:* Azpiazu, D., Gutman, G.E. and Vispo, A. eds. *La desregulacion de los mercados: paradigmas e inequidades de las politicas del neoliberalismo.* Grupo Editorial Norma, Buenos Aires, 34-162.

Gutman, G.E., 2002. Impact of the rapid rise of supermarkets on dairy products systems in Argentina. *Development Policy Review,* 20 (4), 409-427.

Henson, S. and Loader, R., 2001. Barriers to agricultural exports from developing countries: the role of sanitary and phytosanitary requirements. *World Development,* 29 (1), 85-102.

Hu, D., Reardon, T., Rozelle, S., et al., 2004. The emergence of supermarkets with Chinese characteristics: challenges and opportunities for China's agricultural development. *Development Policy Review,* 22 (5), 557-586.

Hueth, B., Ligon, E., Wolf, S., et al., 1999. Incentive instruments in fruits and vegetables contracts: input control, monitoring, measurements and price risk. *Review of Agricultural Economics,* 21 (2), 374-389.

Jank, M.S., Farina, E.M.M.Q. and Galan, V.B., 1999a. *O agribusiness do leite no Brasil.* Editora Milkbizz, Sao Paulo.

Jank, M.S., Leme, M.F.P., Nassar, A.M., et al., 1999b. Concentration and internationalization of Brazilian agribusiness exporters. *International Food and Agribusiness Management Review,* 2 (3/4), 359-374.

Jones, E. and Hill, L.D., 1994. Re-engineering marketing policies in food and agriculture: issues and alternatives for grain grading policies. *In:* Padberg, D.I. ed. *Re-engineering marketing policies for food and agriculture.* Texas A&M, 119-129. Food and Agricultural Marketing Consortium, FAMC no. 94-1.

Mazzocco, M.A., 1996. HACCP as a business management tool. *American Journal of Agricultural Economics,* 78 (3), 770-774.

Milicevic, X., Berdegué, J. and Reardon, T., 1998. *Impacts on rural farm and nonfarm incomes of contractual links between agroindustrial firms and farms: te case of tomatoes in Chile: proceedings of the Meetings of the Association of Farming Systems Research and Extension (AFSRE), November 30-December 4 1998, Pretoria, South Africa.*

Neven, D. and Reardon, T., 2004. The rise of Kenyan supermarkets and the evolution of their horticulture product procurement systems. *Development Policy Review,* 22 (6), 669-699.

Northen, J. and Henson, S., 1999. *Communicating credence attributes in the supply chain: the role of trust and effects on firms' transactions costs: paper presented at the IAMA World Food and Agribusiness Forum, June 13-14, Florence.*

Reardon, T. and Berdegué, J.A., 2002. The rapid rise of supermarkets in Latin America: challenges and opportunities for development. *Development Policy Review,* 20 (4), 371-388.

Reardon, T., Codron, J-M., Busch, L., et al., 1999. Global change in agrifood grades and standards: agribusiness strategic responses in developing countries. *International Food and Agribusiness Management Review,* 2 (3/4), 421-435. [http://www.ifama.org/nonmember/OpenIFAMR/Articles/v2i3-4/421-435.pdf]

Reardon, T. and Farina, E.M.M.Q., 2001. The rise of private food quality and safety standards: illustrations from Brazil. *International Food and Agricultural Management Review,* 4 (4), 413-421.

Reardon, T. and Timmer, C.P., in press. Transformation of markets for agricultural output in developing countries since 1950: how has thinking changed? *In:* Evenson, R.E., Pingali, P. and Schultz, T.P. eds. *Handbook of agricultural economics. Vol. 3: Agricultural development: farmers, farm production and farm markets.* Elsevier, Amsterdam.

Reardon, T., Timmer, C.P., Barrett, C.B., et al., 2003a. The rise of supermarkets in Africa, Asia, and Latin America. *American Journal of Agricultural Economics,* 85 (5), 1140-1146.

Reardon, T., Vrabec, G., Karakas, D., et al., 2003b. *The rapid rise of supermarkets in Croatia: implications for farm sector development and agribusiness competitiveness programs: report to USAID.* Development Alternatives Inc., Bethesda.

Schejtman, A., 1998. Agroindustria y pequeña agricultura: experiencias y opciones de transformación. *In: Agroindustria y pequeña agricultura: vínculos, potencialidades y oportunidades comerciales.* Naciones Unidas, Santiago de Chile. [http://www.fao.org/Regional/LAmerica/prior/desrural/10041.pdf]

Stephenson, S.M., 1997. *Standards and conformity assessment as nontariff barriers to trade.* The World Bank, Washington. Policy Research Working Paper no. 1826.

UNCTAD, 2001. *World investment report 2001: promoting linkages.* UN Conference on Trade and Development, New York. [http://www.unctad.org/en/docs/wir01full.en.pdf]

Unnevehr, L.J. and Jensen, H.H., 1999. The economic implications of using HACCP as a food safety regulatory standard. *Food Policy,* 24 (6), 625-635.

Weatherspoon, D.D. and Reardon, T., 2003. The rise of supermarkets in Africa: implications for agrifood systems and the rural poor. *Development Policy Review,* 21 (3), 333-355.

Zylbersztajn, D. and Neves, M.F., 1997. Illycaffe: coordenacao em busca de qualidade. *In:* Farina, E.M.M.Q. ed. *Estudos de caso em agribusiness.* Biblioteca Pioneira de Administracao e Negocios, Sao Paulo, 47-64.

CHAPTER 9

AGRO-FOOD CHAINS AND SUSTAINABLE LIVELIHOOD

A case study of cassava marketing in Nigeria

OLUSOLA BANDELE OYEWOLE AND BIOLA PHILLIP

Research and Development Centre (RESDEC)
University of Agriculture
P.M.B. 2240, Abeokuta, Nigeria
E-mail: solaoyew@hotmail.com; oyewoleb@yahoo.com

Abstract. Cassava is a good example of problems that local producers encounter in developing agro-food chains aiming at added value and fair rewards for labour inputs. Low production at many small-scale farms lead to high transaction costs. Cassava spoils easily and is costly to transport in its raw form as it consists mainly of water. Therefore much processing takes place on-farm. Processing results in *Gari*, *Lafun* and *Fufu* products with longer shelf life than cassava roots. These products are consumed in the household or sold in the local market. Middlemen buy these products to sell them to urban or international consumers. The products can also serve as basis for further industrial processing but this option is under-exploited so far. It is a highly competitive market with fairly uniform products priced according to the demand–supply principle. Formal quality control is missing. The largest share of added value goes to secondary processors and middlemen. Organizing farmers and training them in entrepreneurship skills is needed to improve their bargaining position and their production and processing process. Policy should provide an enabling environment in terms of banking facilities, quality regulation and control, etc., to support the entire chain. It can support increase in scales of processing at farmers' level, increase in investment in the chain, and promote closer and more sustainable interaction between producers, processors, salesmen and consumers in an agro-food chain.
Keywords: local foods; market structure; banking, labour; processing; organization

INTRODUCTION

Agricultural production in Africa is not very productive per unit of land and unit of labour. It is constrained by lack of access to land, poor technology and harsh environments in terms of low soil fertility, erratic rainfall and fragile ecosystems. On top of that it suffers from marketing constraints. Developing an agricultural produce that is attractive for local or international markets requires quality:price ratios that are competitive. Low population densities are not conducive for extensive

R. Ruben, M. Slingerland and H. Nijhoff (eds.), Agro-food Chains and Networks for Development, 107-115.

infrastructure in terms of roads, transport possibilities, financial services and provision of affordable inputs at producers' level etc. It is very difficult to achieve economies of scale. Transaction costs of agricultural produce are high. Importing produce from Europe into African urban centres may be cheaper than producing locally and transporting over large distances to these centres.

The majority of African smallholders cultivate less than two hectares of farm lands, use rudimentary tools and lack access to processing machines. Lack of appropriate storage facilities prohibits to a great extent the ability to accumulate produce till it becomes a batch that is sufficiently large and attractive to enter a supply chain. Many of these farmers lack organization in producers' organizations and product quality evaluation and are therefore difficult to include in international food chains. In the international context much agricultural produce from developing countries finds it difficult to enter today's international market due to the set quality requirements. Agricultural produce largely remains on farm or goes to nearby local markets.

To provide insight into the problems confronting agricultural product marketing and food chains in Africa the example of cassava marketing in Nigeria is used. Special attention is given to street foods and local market operators.

CASSAVA CASE

Cassava (*Manihot esculenta Crantz*) is a very important crop in Africa and has recently become "the most important root crop in Nigeria" (Ugwu 1996). African countries produce over 81 million tons per year and Nigeria accounts for 45 million tons. Over 70% of production in Nigeria is consumed locally. Nigerian cassava production is by far the largest in the world; a third more than production in Brazil and almost double the production in Thailand and Indonesia (Phillip et al. 2004). However, Nigeria takes a small proportion, about 0.001%, of the world export market. Thailand is a big player accounting for 50% of the EU market. Price differences between 104 US dollars/ton in Nigeria against 21 USD/ton in Thailand reflect the difficulties in production and marketing circumstances in Africa.

Cassava is propagated by stem cuttings and thrives in fairly bad weather and poor soils with little or no fertilizer application. It can be harvested from 6 months to 3 years after planting and the roots can remain in the soil after maturity for more than six months before harvesting. In Nigeria cassava is cropped sole or in association (intercrop) with maize and vegetables. Cassava producers in Nigeria are small-scale farmers that number in their millions. Cassava is available all year round although the labour requirement for uprooting in the dry season is more than during the wet season. There has been a steady growth in cassava production in Nigeria from 12 million tons in 1986 to 31 million tons in 1996 with current production estimated at 34 million tons. This increase is fully due to an increase in number of hectares under cultivation. Average production per ha remained stable at about 11 tons (FMANR 1997).

Cassava forms a major part of the dietary intake of Nigerians, especially in southern Nigeria, and is said to have a daily per-capita dietary calorie equivalent of

238 kcal (Ugwu 1996). A few varieties of the species are eaten boiled but the bulk of the production is processed before utilization. This means that cassava root is not available in food stores or supermarkets. The most popular traditional processed consumer products from cassava include *Gari* (toasted granules), *Lafun*, *Fufu* and starch, while the semi-processed producer goods, i.e., industrial raw materials, are cassava flour, cassava pellets, tapioca, animal feed and industrial starch (UNIFEM 1989).

Producers of the consumer goods are farm families in most cases, and small/medium-scale entrepreneurs. Most farm families engage in cassava processing, firstly, as a way of providing food for the immediate use of the household, and secondly, to add value to the product in order to increase their farm income. Thirdly, cassava is bulky and unlike yam or other root crops it is not directly edible. The crop exhibits certain characteristics in terms of deterioration in quality of the produce if it is not immediately utilized. Its processing extends the shelf-life thereby reducing the risk of wastage and expensive cost of transportation over long distances. The focus of this paper is the marketing of cassava and its effects on the livelihood of the producers.

Structure of the market

Cassava marketing in Nigeria is a model of a competitive market and depicts the following characteristics. The operators are independent and decentralized in decision-making. They have fairly homogeneous products though some exhibit certain levels of price differentiation that Ikpi (2002) reported as having monopolistic tendencies. The general outlook is that the degree of competition in the market is fairly high; hence, we can safely describe the market as belonging to the perfectly competitive industry. Moreover, there exists free mobility of resources in the industry, and buyers and producers are well informed about the industry's activities. The focus of our analysis is on the number and size distribution of the buyers and sellers in the market, the product differentiation and conditions of entry. These variables are important determinants of the magnitude of power the operators have relative to others in the industry.

Cassava producers in Nigeria are independent. They are not unionized, neither do they have agencies that exert any form of control over the producers or marketers. Most of them are small-scale producers located in the rural areas of the country but predominantly south of the river Niger. There exist local and improved varieties of cassava and they differ mainly in terms of yield per hectare, resistance to pests and diseases and maturity dates. The products too are fairly homogeneous and very little attention is paid to coloration, sorting/selecting and even packaging. Buyers are equally large and have no forum to discuss or agree on prices. In essence, therefore, the principle of demand and supply is the key to pricing in this market. As cassava comes from many small units, each individual farmer has very little control over prices. The high dependency on weather and biological patterns of production implies that marketing agencies in the short run must adjust to farm supplies.

The cassava food chain

As an agricultural product, cassava is largely a raw material for further processing. The product soon loses identity and becomes food. It is bulky, and this single characteristic also has implications for physical handling in terms of haulage, that is, transportation cost, storage space and risk. It also has the tendency to have reduced quality if not processed soon after harvesting. All these have a lot of implications for the facilities necessary to market the crop at different stages.

Cassava is usually processed as follows. First the outer coating (cassava peel) is removed. The whitish part left is then thoroughly rinsed (washed) before the processing for *Gari, Lafun, Fufu* or other products starts. Many food products made from cassava in Africa are products of fermentation (Oyewole and Odunfa 1992). The duration and method of fermentation vary depending on the product under consideration. The fermentation processes serve to reduce the cyanide content of cassava and to impart palatability to the product and they also increase shelf life (UNIFEM 1989).

Fermented cassava flour (*Lafun*) is usually made from freshly harvested cassava roots. The roots are peeled and subjected to a fermentation and drying process. The drying process helps to increase the shelf life and reduces the bulkiness of the product. Milled dried fermented cassava root materials give the *Lafun* (cassava flour).

To produce *Gari* the peeled cassava is grated and the pulp is bagged and compressed to express the water while undergoing fermentation. The dewatered pulp is sieved and roasted. This reduces the bulk and weight and increases the shelf life. A well processed *Gari* can be stored for two years without adding preservatives. This product is easily transported to urban markets several kilometres away or as export commodity.

Two forms of *Fufu* are traditionally produced in Nigeria; wet *Fufu* paste and ready-to-eat *Fufu*. The third form is a recently produced *Fufu* powder. The peeled cassava is usually immersed in water to ferment. The water is pressed out and the pulp is pounded, wrapped firmly in leaves or nylon and steamed. The later processing method stops at the pounding, i.e steaming is not done. This means that the consumer would have to steam before serving. This is commonly transported to urban centres while the former is usually sold a few kilometres away from the point of processing. The shelf life of both forms of *Fufu* is about 9 days. The third method is the outcome of a recently concluded research work (Oyewole et al. 2001). *Fufu* powder has been test-marketed and is currently undergoing widespread publicity, large-scale production and commercialization.

These products (*Fufu, Gari* and *Lafun*) are processed mostly by the farmers themselves, and also by middlemen who buy fresh roots and process them into any of the products above. There are a few medium- or large-scale producers of *Gari*, but the bulk of *Lafun* and *Fufu* producers are small-scale. Though expensive to produce because it is labour-intensive and requires a high fuel-wood consumption, a high percentage of cassava roots produced in the forest and savanna regions of Nigeria are processed into *Gari* (Nweke 1994). There is no formal quality control on cassava-based products or processing methods, neither with respect to nutritional quality nor on hygiene or other characteristics related to food safety.

The chain of marketing cassava products is indicated in Figure 1.

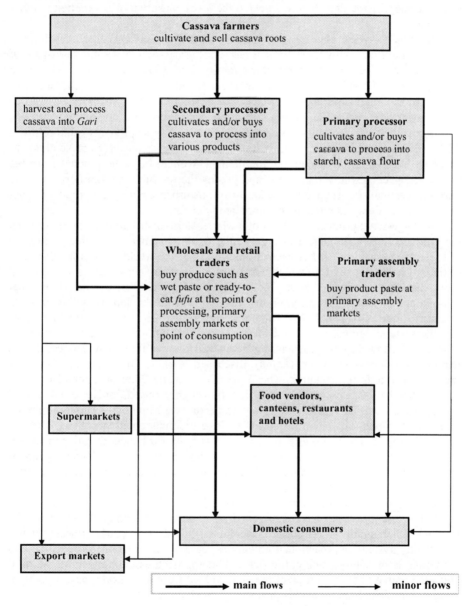

Figure 1. Typified marketing chains for cassava in Southwest Nigeria (modified after Dipeolu et al. 2001)

The primary processors are mainly farmers. The middlemen enter the chain as wholesalers and or retailers. Typically, the cassava roots are harvested and processed close to the point of harvest. The secondary processor adds value to it and then sells to wholesalers or retailers, before it gets to the local consumer or is exported to other countries.

The link between the cassava growers and the consumers of the finished products could be relatively short at times. As the cassava leaves the farm gate, it could be processed by the same farm family into *Fufu, Lafun* or *Gari*. These products could be sold in the village markets (or nearest market) by the producer herself to the final consumer. It is also possible for such products to be sold to middlemen. The middleman in turn will move such produce to the urban centre for retailing to the final consumer. At other times, some of these products could follow a fairly longer chain, e.g. *Gari*, as it travels over long distances from southern to northern Nigeria. In this chain, the middleman/-woman buys the product from the producers, packages it in bags, sorts and grades the product and transports it to the northern destination. At specific market locations/urban centres, the product is then sold to wholesalers and retailers to complete the chain.

The large-scale producers sell through wholesalers (distributors), who in turn sell the products to retailers before they get to the final consumer. A sizeable proportion of these are transported to Europe and other parts of the world to meet the demands of the Nigerian-cum-African population. Marketing of cassava products as industrial raw materials has remained largely unexploited in Nigeria. Research has shown that cassava can be substituted for grain in livestock feed production, as is done in parts of Latin America (Ugwu 1996). There are unexploited opportunities for the export of pellets, starch, glucose syrup and alcohol from cassava. Post-harvest utilization of cassava has moved to the production of industrial cassava flour. This process differs from that of cassava flour (*Lafun*) discussed earlier. Industrial cassava-flour production avoids fermentation. The processing can be done on a small or large scale. After harvest, the cassava is peeled, thoroughly rinsed, milled, dried, sieved and packaged. All harvested cassava has to be processed within 24 hours to obtain high-quality flour. The chain in industrial flour products could stop here, whereby the product is sold to factories that will use it as input for bakers (bread, pancake or biscuit), or the processing can continue for the production of glucose syrup or ethanol. These products are in turn sold to confectionaries, pharmaceutical industries and producers of alcoholic beverages (spirits). The final products are then sold to wholesalers and retailers to complete the chain (Figure 2).

A few organizations have intervened in cassava marketing. A report of the International Institute for Tropical Agriculture (Phillip et al. 2004) indicated that a commodity-system approach that integrates cassava production, processing and marketing is presently being put in place. The approach assists farmers in the rural areas to retain a high proportion of the value-added from processing cassava to high-quality flour. The final product is then sold to the baking industry.

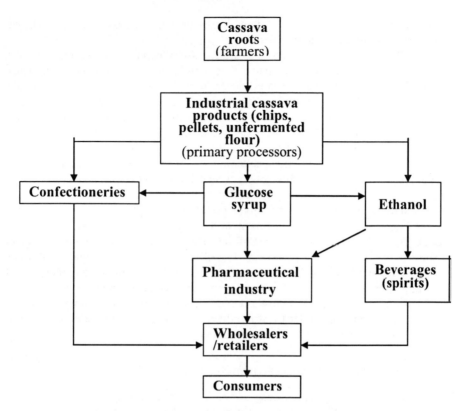

Figure 2. Typical market chains of cassava industrial products

Sustainable livelihood and economic development

The essence of marketing cassava-based products is to receive value for labour. At every stage in the cassava-processing chain, labour and materials are used up to add value to the product. The longer the chain, the larger the number of people involved in the process. Women constitute the major source of labour for cassava processing and marketing. They often buy cassava in the soil, they harvest, process and market. This increased earning opportunity enables them to purchase goods and services that contribute positively to their livelihood. According to Dipeolu et al. (2001), cassava generates the largest income for the largest number of farming households in Nigeria. About 34% of total household farm income in Ogun and Imo states and about 20% in Benue state are generated annually from cassava-related activities. The retailers of the *Lafun*, *Fufu* and *Gari*, the restaurant owners and the street food vendors who are involved in taking the finished product to the consumer also benefit in terms of better livelihoods. The benefits of wholesalers range from 20 to 50% and those for processors/retailers between 50 and 70% (Dipeolu et al. 2001). The farmers, who provide most of the labour in the production chain, do not make as much profit as other actors in the chain.

The extension of the markets that derive from cassava processing and the marketing chain is an important part of economic development and sustainable livelihood. The markets enable the producer to exchange his products for income (credit/cash) which can be utilized to improve livelihood. Cassava processing provides employment to producers, transporters, processors, marketers and food vendors (Phillip et al. 2004). The local cassava-marketing chain has aided the development of product marketing. There have been noticeable improvements in marketing segments like grading, packaging (to enhance product durability) and storage. These resulted in quality products that attract better prices and enhanced income.

The continuous growth of the urban population coupled with some policies embarked upon by the Nigerian government has spurred up the demand for cassava products especially since the later part of the 1980s. To this extent cassava products are sources of food of great preference to rich and poor in urban and rural areas alike. An effective local cassava market chain fits to these growing demands. It has at the same time been a source of foreign-exchange savings for the nation as a whole (less food import needed). Moreover, it has earned the prominent position of the 'poverty alleviation crop' in terms of the diverse roles it plays in the economic life of the Nigerian economy.

Constraints to optimal utilization of cassava in Nigeria

While there exists a large opportunity for cassava producers to earn good incomes and live comfortably, there is still a lot of poverty. The odds against maximizing the utility of cassava in Nigeria are many.

A sizeable number of the producers operate on a very small scale that could be considered economically non-viable. Their inability to analyse effectively the cost–benefit returns from the activity or a lack of proper machinery to dispose of the cassava produced might be the only reason why they are still engaged in the business. To this extent, if demand for cassava roots by large-scale firms exists, such farmers might rather sell their produce. For the moment the absence of large-scale firms that may take advantage of the economies of large-scale production further hinders progress in the cassava-producing chain.

Many producing areas still lack good communication networks. The bulk of the producers travel short distances to the nearest market (urban or rural) to dispose of their produce. Cost of transportation is still high and further reduces the profit margin of producers.

It has been observed that producers (farms) command a relatively low share of the wholesale and retail price and sometimes sell their produce on credit, at least in the *Fufu* market. Attention must be directed towards proper marketing of cassava products in Nigeria.

More importantly, an effective marketing system for cassava with the aim of promoting development must have a sound policy backing. As it were, the cassava market is an unguided industry with operators behaving as they wish without reference to any guideline. The haphazard nature of the industry is partly responsible for its lack of development. Inadequate funding of research institutions to conduct research and the publication of results on problems identified by small-scale processors hinder coherent development in the industry.

CONCLUSION

What can be done to make cassava food chains an instrument of development? There are currently three major challenges: quality assurance, improved production and processing capacity and overcoming market limitations. In addition one might explore new or unexploited markets for the smallholders. To enter these markets smallholders and street food participants need to be integrated into agro-food chains and networks. When successful this will increase rural livelihoods. Yet, many things are needed to achieve product, process and marketing improvements.

To counteract the scale problems smallholder farmers need to organize themselves. Farmers' organizations such as processing cooperatives can also be effective in creating added value and reaping the benefits from it. To improve actual practices a system of continuous informal education is needed, especially training on quality assurance and marketing. A profitable cassava supply chain needs an enabling environment consisting of infrastructural support such as agricultural banking, agricultural insurance schemes, export promotion boards, etc. Regional partnership and international cooperation can be of assistance in shaping the circumstances for effective cassava supply chains and networks providing safe and nutritious products for local, regional or international markets.

REFERENCES

Dipeolu, A.O., Adebayo, K., Ayinde, I.A., et al., 2001. *Fufu marketing systems in South-West Nigeria.* Natural Resources Institute, University of Greenwich, Chatham Maritime. NRI Report no. R2626. [http://www.nri.org/research/rootcropsA0898.doc]

FMANR, 1997. *Yearly statistics of agricultural production in Nigeria.* Federal Ministry of Agriculture and Natural Resources, Abuja.

Ikpi, A., 2002. *Policy directions and performs for a competitive Cassava subsector in Nigeria: invited paper for a workshop on "Cassava Competitiveness", Ibadan, Nigeria, November 18-22.*

Nweke, F.I., 1994. Cassava processing in Sub-Saharan Africa: the implications for expanding cassava production. *Outlook on Agriculture,* 23 (3), 197-205.

Oyewole, O.B. and Odunfa, S.A., 1992. Effect of processing variables on cassava fermentation for "fufu" production. *Tropical Science,* 32, 231-240.

Oyewole, O.B., Sanni, L.O., Dipeolu, A.O., et al., 2001. *Factors influencing the quality of Nigerian fufu: presented at the 8th triennial symposium of the International Society for Tropical root crops - Africa Branch, 12-16 November.*

Phillip, T.P., Sanni, L.O. and Akoroda, M., 2004. *A cassava industrial revolution in Nigeria.: the potential for a new industrial crop.* International Institute for Tropical Agriculture, Ibadan.

Ugwu, B., 1996. Increasing Cassava production in Nigeria and prospect for sustaining the trend. *Outlook on Agriculture,* 25 (3), 179-185.

UNIFEM, 1989. *Root crop processing.* United Nations Development Fund for Women, New York. Food Cycle Technology Source Book no. 5.

BUSINESS CASES

CHAPTER 10

SUPPLY-CHAIN DEVELOPMENT FOR FRESH FRUITS AND VEGETABLES IN THAILAND

JAN BUURMA[#] AND JOOMPOL SARANARK[##]

[#] LEI, Wageningen University and Research Centre, P.O. Box 29703, 2502 LS,
Den Haag, The Netherlands. E-mail: jan.buurma@wur.nl
[##] Department of Agriculture, Thailand

Abstract. In this case study two supply-chain development projects in Thailand are analysed:
1. TOPS Thailand: introduction of food safety standards for the domestic market.
2. Fresh Partners: development of an integrated quality chain for the export market.
TOPS Thailand is a retail company with about 50 supermarkets in Bangkok and Chiangmai. The management decided to introduce a certification system for food safety in order to improve their competitive position and to consolidate their image of a quality supermarket. The introduction resulted in a system of preferred suppliers that had to obtain a certificate for good agricultural practices from the Department of Agriculture. The number of suppliers sharply dropped in the course of the project period. Fresh Partners Thailand is an export company shipping exotic vegetables from Thailand to The Netherlands and surrounding countries in Europe. The management decided to develop an integrated quality chain in order to comply with the increasing food safety requirements in the European Union and Japan. The investments in quality systems coincide with a growing demand for exotic vegetables in north-western Europe. Consequently export volumes and numbers of smallholders and labourers are rapidly growing.
Keywords: food safety; export; retail; good agricultural practices; quality systems

INTRODUCTION

Since 1999 researchers of the Agricultural Economics Research Institute LEI – a part of Wageningen University and Research Centre (Wageningen UR) – were involved in the implementation of two completely different supply-chain development projects in Thailand. One project was focused at the domestic market and the other at the export market. In this case study the experiences and findings gathered during project implementation are set side by side. The exercise gives a highly interesting glimpse behind the scenes of supply-chain development. The first project represents the conduct of a retail company (TOPS Thailand) and the second the conduct of an export company (Thai Fresh) in supply-chain development. A retail company occupies an other position between producer and consumer than an export company. Consequently their strategies with regard to supply-chain

R. Ruben, M. Slingerland and H. Nijhoff (eds.), Agro-food Chains and Networks for Development, 119-127.

development may be different. As a result the impacts for smallholder development and sustainability may also diverge.

The ambition of the case study is to identify both critical success factors and critical success actors for supply-chain development. Starting supply-chain development from a retail company has other effects for smallholder involvement and sustainability than starting from an export company. Policymakers and business partners should be aware of these effects and include them in the strategic decision-making process for supply-chain development.

TOPS THAILAND PROJECT

In 1996 the Dutch retail company Royal Ahold established a joint venture with the Central Retail Corporation in Thailand, running over thirty TOPS supermarkets in Bangkok and Chiangmai. The management was instructed to transform TOPS into a flourishing high-quality supermarket chain. As a first step World Fresh, the distribution centre for fresh products, was established. Furthermore, the product flow from the distribution centre to the individual branches was streamlined. Meanwhile the economic recession in Asia also affected Thailand. Consequently the TOPS management had to undertake actions to improve their competitive position. Cost reduction and quality improvement at the upstream side of the distribution centre became the strategic attention points. At this point researchers of Wageningen UR were enlisted to elaborate options for implementation. The process was supervised by a Steering Committee, consisting of executives of companies and institutions, directly or indirectly involved in supply-chain development.

Technical experts quantified the possible reductions in transaction costs of bringing down the number of suppliers for individual fresh products. These calculations resulted in a strategy of preferred suppliers. Subsequently socioeconomic experts were enlisted to elaborate the strategy of preferred suppliers and simultaneously safeguard product quality with regard to food hygiene and pesticide residues. The final outcome was that preferred suppliers had to operate under a certification system for good agricultural practices. The TOPS management decided to embrace the certification system for good agricultural practices of the Department of Agriculture (DoA). The suppliers were bound to obtain a certificate from DoA.

The pathway to certification included two phases. In the first phase the socioeconomic experts identified five actual production systems for vegetables in Thailand (Table 1) and suggested to give preference to growers presently applying the production system of 'Intelligent Pesticide Management'. The reasons for this preference were threefold: balanced use of fertilizers and pesticides, readiness to comply with certification standards, and enough production capacity to safeguard a continuous supply of fresh vegetables. The information in Table X.1 makes clear that the choice for 'Intelligent Pesticide Management' leads away from the smallholders who are traditionally supported by public or semi-public institutions like agricultural extension, government-supported projects and non-governmental organizations.

Table 1. *Qualifications of five production systems for vegetables in Thailand according to the use of agro-chemicals, development context and certification framework*

Characteristic	Conventional local-market growers	Conventional professional growers	'Intelligent Pesticide Management'	Integrated pest management	Organic
Use of synthetic pesticides	High	High	Reduced	Low	None
Use of artificial fertilizers	Divergent	Optimal	Balanced	Balanced	None
Institutional support	Agricultural extension	Input uppliers	Input suppliers	FAO-project Non-Form.Ed	Various NGOs
Development approach	Top-down	Participative	Participative	Bottom-up	Bottom-up
Development objective	Technology application	Yield security	Save product	Pest prevention	Sustainable agriculture
Development phase	Struggling	Standing	Arising	Pioneering	Pioneering
Certification standard	None	None	FAO codex	FAO codex	IFOAM
Certification level	None	None	Product	Process and product	Process
Certification agency	None	None	Agricultural departments	Agricultural departments	Still lacking
Certification label	None	None	Non-toxic	Non-toxic	Organic
Residue analysis	Public health	Supermarket	Agricultural departments	Agricultural departments	Not relevant

In the second phase the socioeconomic experts checked the opinions on certification among the various stakeholders within and around the supply chain. The majority of the stakeholders (8 out of 10) were at least conditionally positive on certification (Table 2). The information of Table 2 makes clear that (unfortunately) the vegetable brokers and the buying department were negative about certification. They wanted to keep their hands free for transactions with non-certified partners. Furthermore the costs of certification gave rise to long discussions. Finally the TOPS management obliged all suppliers of fresh vegetables and fruits to obtain a certificate from the Department of Agriculture, thus bypassing the objections of vegetable brokers and the buying department.

Due to financial problems Royal Ahold was forced to discontinue their participation in the TOPS supermarkets in Bangkok. The involvement of researchers of Wageningen UR was also discontinued. According to recent information the new owners of TOPS have continued the certification relationship with the Department of Agriculture.

Table 2. Opinions on certification of stakeholders involved in the supply chain of fresh vegetables and fruit in Thailand

Stakeholder	Opinion
Crop-protection associations	positive
Pesticide companies	conditionally positive
Seed companies	conditionally positive
Vegetable growers	conditionally positive
Vegetable brokers	negative
Buying department	negative
TOPS / World Fresh	positive
Consumer-interest groups	positive
Inspection bodies	conditionally positive
Dept. of Agriculture	conditionally positive

Lessons learned

The TOPS management had to operate under rather difficult business conditions. They decided to focus on reduction of transaction costs and improvement of food safety levels. As a result a selection process among the original suppliers was initiated. The more professional and advanced growers and traders achieved a preferred position. Their less professional and advanced colleagues had to abandon the field. Consequently the integration of smallholders in the supply chain of TOPS was reduced. The decision to select growers already applying 'Intelligent Pesticide Management' implies a kind of disqualification of the public and semi-public institutions (like agricultural extension, government supported projects and NGOs) that traditionally support smallholders. The Department of Agriculture has operated quite visionary by developing a certification system for good agricultural practices. On the other hand DoA has been manoeuvred into a vulnerable position. Retailers can hide themselves behind DoA when 'certified' products at some time turn out to be substandard. In such cases the Ministry of Agriculture may suffer a loss of face.

Vegetable traders and the buying department felt themselves restricted in their freedom of transaction by the requirements of certification. Certification makes it more difficult to take refuge to cheap solutions or to occasional suppliers. For businessmen the job satisfaction is often found in this type of opportunities. This means that private and public policymakers should not count too much on the cooperation of businessmen in certification processes.

The socioeconomic experts of Wageningen UR got easy access to both public and private parties in and around the fresh vegetable and fruit supply chain in Thailand. They further got the impression that contacts between public parties and private parties were exceptional. This means that university researchers can play a very constructive role in supply-chain development as mediators between public parties and private parties.

Stages and conditions

During the project period different stages followed each other. In each stage different basic conditions for growth were actual. The stages and conditions are specified in Table 3.

Table 3. Successive stages and matching conditions for growth in TOPS Thailand project

Period	Stage	Basic conditions for growth
1999	Forming	Competition
2000	Organizing	Risk/return
2002	Implementing	Government involvement

THAI FRESH PROJECT

The Thai Fresh project was initiated in 1999 when Golden Exotics Holland and KLM Cargo established a distribution and packing centre in the vicinity of Bangkok airport. In the years before, Golden Exotics had already built up a good reputation in the distribution of exotic vegetables from Thailand in Germany, United Kingdom, The Netherlands and Belgium. In those years fresh products were purchased from wholesalers and brokers. This mode of sourcing was no longer workable, owing to the increasing quality and safety requirements of the international end-markets in the EU and Japan. In fact Golden Exotics Holland faced increasing problems with the Dutch Inspectorate for Health Protection regarding pesticide residues. From 2002 on, researchers of LEI were actively involved in the project. The involvement of LEI was co-funded by the Dutch Ministry of Development Cooperation (SENTER – PSOM programme).

The Thai Fresh project aimed at the development of an integrated quality chain for export of exotic vegetables. The challenge of developing such an integrated quality chain is translating the quality and safety requirements at retail level into good agricultural practices at producer level and to develop a supply-chain structure for a reliable tracing and tracking system. The challenges concerned were tackled in two successive actions: (1) the establishment of a distribution and packing centre at Bangkok airport, and (2) the establishment of a regional post-harvest centre in Ratchaburi province.

The establishment of the distribution and packing centre at Bangkok airport was a first step in getting a better control on product quality and food safety. In the beginning the fresh products were purchased from Bangkok-based wholesale traders. After delivery at the distribution and packing centre the products are graded, sorted, washed, packed and temporarily stored in a cold room, where pallet build-up for freighting, inspections by customs and the quarantine service are executed in the meantime. The distribution and packing centre can be regarded as value-added centre, where grades and standards are implemented and where compliance with these standards is enforced. HACCP has been introduced at the distribution and packing centre in order to arrive at good manufacturing practices (GMP).

Introduction was accompanied by the development of Standard Operating Procedures (SOPs) and the implementation of a training programme for the managers and the workers at the centre.

The establishment of the distribution and packing centre was prompted by developments in the international end-markets in the EU and Japan. In the late 1990s consumer confidence in EU and Japan reached an all-time low. Consumers began to demand more transparency in the food chain. This transparency included the verification of the composition of the product, its origin and traceability, its safety, and the claims that were made concerning product features like nutritional values, health effects, etc.

Sourcing from Bangkok-based wholesale traders implied a number of weaknesses regarding quality and safety assurance:

- The lack of quality control at the farm led to a variable quality of vegetables. Subsequently, this resulted in a relatively high level of rejection of substandard quality at export destination and hence financial loss due to waste.
- The fact that there was no recognized standard of quality in Thailand also resulted in a decreasing access to the EU markets and prevented new access to the high-value Japanese market.

The distribution and packing centre in Bangkok was not sufficient to solve these weaknesses. Therefore a further upward integration of the supply chain was considered to be necessary. For that purpose a regional post-harvest centre was built (2003) in the production region of Ratchaburi province. The post-harvest centre also serves as a knowledge centre for the growers. The centre provides the growers with extension services and farming inputs so that they can apply good agricultural practices and integrated crop management techniques. The services concerned have the target to get the growers certified according to EUREP-GAP.

Pesticide residues are the most important food safety concern in the vegetable supply chain. For that reason farmers are being trained in good agricultural practices (GAP) with regard to pesticide application. The GAP terms of reference imply: (a) minimizing the use of agrochemicals and implementing a traceability system; (b) becoming aware and taking care of environmental protection and efficient use of resources; (c) assure the workers' health, safety and welfare. The training is provided by a team of experts consisting of a full-time extension worker of Thai Fresh, an agronomist From Kasetsart University and back-up support from Bureau Veritas (certification company) and LEI. The training includes the preparation of a pesticide policy manual for the contract growers and assistance in setting up a record-keeping system.

The establishment of the regional post-harvest centre in Ratchaburi implies a shortening of the supply chain by by-passing the wholesale traders. As for the primary production level, commitment from the growers is created through contractual agreements on purchases and by making them shareholders of the regional post-harvest centre.

The organizational structure of the Thai Fresh supply chain has been depicted in Figure 1.

Thai Fresh supply-chain redesign

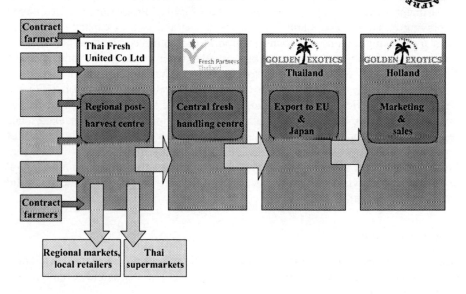

Figure 1. Organizational structure of the Thai Fresh integrated quality chain

The Thai Fresh Business model (Figure 1), combined with the strategy for building competences along the various levels of the supply chain, appears to be quite successful. The export volumes to Europe show an impressive growth rate. The present export results are considered so promising that export to Japan has no priority for the time being.

In the near future a demonstration garden on the land adjacent to the post-harvest centre will be established to support the suppliers/growers further. Follow-up investments in Thailand and Vietnam are under consideration. The focus is on application of the business model at other products and other countries.

Lessons learned

The Thai Fresh management is operating in the rapidly growing market of exotic vegetables in Europe. Such a position makes investments in buildings, certification systems and human resources easier. The management decided to make such investments in order to maintain their access to the market of exotic vegetables in Europe. In fact the increasing need for food safety assurance was the driving force to develop an integrated quality chain. The strategy for building competences along the various levels of the supply chain has enabled numerous smallholder growers to link up with international standards. Simultaneously the involvement of wholesale traders has been reduced. The high priority for building competences may result from the professional background of the Thai Fresh president. His career started in education.

The implementation of the grades and also the auditing and inspection of compliance has been completely in private hands. Government agencies like the Department of Agriculture and the Department of Export Promotion were hampered in playing an active role. Limited financial means and lack of capacity restrained them from taking a more prominent position in promoting commercial horticulture. The communication between public and private parties in both Thailand and The Netherlands is still sub-optimal. The specific reasons for interventions by the Dutch Inspectorate for Health Protection are not communicated to Thai Fresh or to the Department of Agriculture in Thailand. Researchers of LEI got access to all parties involved and assembled a Thai-Dutch dialogue on food safety. This experience again shows the potential of researchers in bringing public and private parties together in dialogues, seminars, lectures, etc. The most crucial element in the Thai Fresh quality chain seems to be the provision of inputs (basis for good agricultural practices) and provision of market access (both international and domestic; solid basis for commitment).

Stages and conditions

During the project period different stages followed each other. In each stage different basic conditions for growth were important. The stages and conditions are specified in Table 4.

Table 4. Successive stages and matching conditions for growth in Thai Fresh project

Period	Stage	Basic conditions for growth
2000	Forming	Access to markets – legal
2002	Organizing	Access to markets – institutional
2004	Implementing	Trust
2005	Optimizing	Risk/return

CONCLUSION

In this section the development pathways and the matching results of TOPS Thailand and Thai Fresh are compared. The comparison aims at formulating conclusions or hypotheses with regard to smallholder involvement and the roles and contributions of public agencies, institutions, public–private partnerships and knowledge centres.

The two projects under consideration had to operate under quite different institutional and economic conditions: TOPS Thailand as a retail company in a period of economic recession in Thailand; Thai Fresh as an export company in a period of booming business for exotic vegetables in Europe. Nevertheless, the strategic choices made during the project periods reveal something of the aims and values of both companies. For TOPS Thailand as a retail company competition and risk/return appeared to be the dominant motives for supply-chain development. Improvement of food safety and reduction of transaction costs were the major strategic attention points. As a result many smallholder producers had to abandon the field.

For Thai Fresh as an export company access to markets from both legal and institutional perspectives appeared to be the dominant motives for supply-chain development. In this case introduction of quality systems and building competences were the major strategic attention points. As a result numerous smallholder producers succeeded in linking up with international standards for good agricultural practices. On the other hand wholesale traders were excluded from participation in the international supply chain of exotic vegetables.

Comparison of both cases leads to the hypothesis that supply-chain development around an export company provides better perspectives for smallholder involvement and sustainability than supply-chain development around a retail company.

The public and semi-public agencies which traditionally support smallholders (like agricultural research and extension and NGOs) played just a minor role in supply-chain development. This may be due to their weak positions in both horticulture and social sciences. The two cases in Thailand have shown that social factors like perceptions, values, visions and strategies of stakeholders represent an important dimension in supply-chain development.

University researchers seem to have comparative advantages regarding access to public parties and private parties. Contacts among public parties and private parties appeared to be exceptional in Thailand (and also in other countries). This means that university researchers can play a very constructive role in supply-chain development as mediator between public and private parties in building public–private partnerships.

CHAPTER 11

FRUITFUL

Integrated supply-chain information system for fruit produce between South Africa and The Netherlands

ANNEKE POLDERDIJK[1], ELSBETH VAN DYK[2], DAISY FERREIRA[3], EGBERT GUIS[4] AND SANDRA KELLER[5]

[1] *Agrotechnology and Food Innovations, P.O. Box 17, 6700 AA Wageningen, The Netherlands. E-mail: anneke.polderdijk@wur.nl*
[2] *CSIR Transportek, P.O. Box 320, Stellenbosch 7599, South Africa. E-mail: fevandyk@csir.co.za*
[3] *Capespan Ltd, P.O. Box 505, Bellville 7535, South Africa. E-mail: dawie_ferreira@capespan.co.za*
[4] *TNO Inro, P.O. Box 6041, 2600 JA Delft, The Netherlands. E-mail: g.w.guis@inro.tno.nl*
[5] *PPECB, P.O. Box 15289, Panorama 7506, Capetown, South Africa. E-mail: South Africanndrak@ppecb.com*

Abstract. South-African and Dutch research institutes and business partners collaborated in several pilot projects for improving logistical performance and quality performance to strengthen the market position of South-African fruit after deregulation of the domestic market. Based on participatory problem assessment, it was concluded that major bottlenecks for realizing fully integrated exchange of information were far less of a the technical nature (hardware and software), but had to do with the cost side and the human' nature, e.g., education, procedures, data accuracy, mistrust, competition, institutional capacities and organization. Pilots were conducted to improve inter-company planning, coordination and information exchange, and to enhance the role of the government in moving towards standardization.
Keywords: deregulation; logistics; electronic data exchange; black empowerment; traceability

INTRODUCTION

The FRUITFUL project focused on improvement of information exchange within the refrigerated fresh-fruit supply chain between South Africa and The Netherlands. The hypothesis was that improvement of information exchange would result in improved logistical performance and improved quality performance, which would strengthen the market position of South-African fruit and that of related fruit supply

R. Ruben, M. Slingerland and H. Nijhoff (eds.), Agro-food Chains and Networks for Development, 129-140.
© 2006 *Springer. Printed in the Netherlands.*

chains on the world market. As a result of this, stakeholders in the fruit export supply chain from South Africa to the Netherlands would profit. In addition, through global developments and requirements such as traceability and certification, a more integrated exchange of information within international agro-food chains was required. Failing of compliance with these new requirements would make it increasingly difficult for South-African fruit and the related supply chains to compete on the world market.

It was decided to follow a flexible approach regarding the system design in order to allow a choice between an overarching system that would replace existing systems and a decentralized system focusing on interfaces between existing facilities.

SOUTH-AFRICAN FRUIT EXPORT INDUSTRY

South Africa's climate and soil condition provide ideal conditions for many varieties of fruit to be grown; deciduous, citrus and subtropical fruit are all grown throughout most of the country. Deciduous fruit includes table grapes (grapes grown for eating, not wine production), pome fruit (apples and pears) and stone fruit (apricots, peaches, nectarines and plums). Citrus is split into oranges, grapefruit, lemons, limes and soft citrus (also known as easy peelers, such as naartjies, mandarins, etc.). Subtropicals are mangoes, litchis, melons, avocados and pineapples (while bananas also fall into this category, South Africa does not export any bananas).

A large amount of South-African fruit is exported. The destination markets are Continental Europe (43%), United Kingdom (25%), Middle East (11%), Far East and Asia (8%), Russian Federation (7%), Americas (3%), Africa (2%) and Indian-Ocean Islands (1%) (Fruit South Africa 2004).

South-African fruit is currently exported through the South-African ports of Cape Town, Port Elizabeth, Durban and the port of Maputo in Mozambique. Fruit is exported in cartons on pallets, either in refrigerated containers or in bulk' shipments, i.e., the pallets are loaded into the hold of a specialized refrigerated vessel. Almost no fruit is exported via air freight as this is too expensive. In addition to the container terminals, there are six dedicated fruit terminals (conventional terminals) in total at the four ports for loading specialized refrigerated vessels. Fruit to the European market is for the most part exported through the Cape Town harbour.

CONTEXT OF THE FRUITFUL PROJECT

Deregulation

Before the deregulation of the marketing of agricultural produce in 1997, only a few parties controlled the South-African fruit export industry, e.g., all citrus fruit was exported through *Outspan* and all deciduous fruit was exported through *Unifruco*. After the deregulation *Outspan* and *Unifruco* became *Capespan*. During regulation South Africa had developed a central information system that was very innovative for that period. After deregulation, however, the situation changed dramatically and became rather chaotic (McKenna 2000). Hundreds of new exporters entered the market and many failed within a short time. Producers were generally not familiar with the free-market system and lacked the experience to face the challenges of the new rules and/or possibilities. Consequently, the (uniformity of the) quality of the fruit at the overseas markets decreased. Prices decreased and many producers and exporters alike faced enormous debts and filed bankruptcy. Global developments were most to blame for the state of the fresh-produce industry; however, deregulation was most certainly a contributing factor. The benefits of the original central information system lost part of their value as many companies and organizations began to develop their own system. As a result of the negative aspects of deregulation South-African fruit lost its (high-quality) reputation on the international market.

After a few turbulent years the fruit industry began to improve its organization in order to rebuild trust between the chain partners and to regain its position on the global fresh-produce markets. A pre-feasibility study of the FRUITFUL project (2000) showed that exchange of information strongly needed review, as there was no uniformity.

Black empowerment and the previously disadvantaged

After the legal abolishment of racial discrimination (Apartheid) in South Africa in 1994 programmes were set up for the benefit of the previously disadvantaged. Before the FRUITFUL project commenced, the South-African fruit industry had already become involved in a transformation programme in order to include previously disadvantaged communities in all spheres of the industry (BuaNews 2003). This included a development programme for emerging farmers as well as assistance for black economic empowered groups. It is estimated (2004) that there are presently 6000-7000 emerging farmers in South Africa.

Labour skills and HIV/AIDS

Due to the seasonal nature of the fruit industry much occasional/seasonal labour is employed. Many new workers are employed every year, as many workers do not return to the same farms year after year. These workers need training in the procedures and equipment used. In addition, many workers have a very low level of schooling or none at all.

HIV/AIDS has made and is continuing to claim many victims in South Africa, and this has an impact on production and on both social and economic circumstances. In the fruit industry, the impact of HIV/AIDS has been more noticeable on skilled labour, e.g.,. scanning and entering of information in the packhouse, forklift drivers, workers in conventional fruit and container terminals.

Global developments

Worldwide the fruit supply chain is moving towards a small number of powerful retailers and towards category management. Individual role-players are searching for ways to strengthen their position in the chain, e.g., by developing value-adding services.

Traceability was gaining more attention. In order to comply with EUREP-GAP and the imminent General Food Law (2005), stakeholders in the fruit chains from South Africa to The Netherlands needed to move towards huge changes in certification and traceability.

If the fruit export chains between South Africa and The Netherlands could comply with the requirements, all role-players could benefit from the global developments.

OUTLINE OF THE FRUITFUL PROJECT

The project commenced in August 2001 and continued until April 2003. Table 1 shows the formal participating parties.

Table 1. FRUITFUL formal participants

Name of participant	Type of organization / role in FRUITFUL
Capespan Ltd	South-African fruit exporter
Intertrading Ltd	South-African fruit exporter
South-African marine Ltd	International shipping liner reefer containers
Seatrade Ltd	International shipping liner
Anlin Ltd	South-African shipping agent
Seabrex BV	Dutch fruit terminal
Hagé International BV	Dutch importer of fruits and vegetables
FTK Holland BV	Dutch importer of (sub)tropical fruits
Paltrack Ltd	South-African IT service provider
PPECB	South-African perishable-products export control board
Rotterdam Municipality Port Management	Dutch Port authority
Dutch Ministry of Agriculture, Nature Management and Fisheries	Dutch sponsor
South Africa – Netherlands Transport Forum	Dutch sponsor
KLICT	Dutch main sponsor
Agrotechnology and Food Innovations BV (used to be ATO until October 2003)	Dutch post-harvest agro-research institute
CSIR Transportek	South-African research institute
TNO Inro	Dutch research institute, FRUITFUL project management

Several other companies contributed to FRUITFUL besides the formal project partners. Table 2 gives an overview.

Table 2. Non-formal parties that collaborated with the FRUITFUL team

Name of collaborating party	Type of organization / role in FRUITFUL
Cape Citrus Ltd	South-African exporter of citrus
Cape Reefers Ltd	South-African conventional reefer shipping liner
SOUTH AFRICAN FT Ltd / CFT Container Fruit Terminal	South-African Fruit Terminals
FPT Ltd	South-African Fresh-Produce Terminals
Coolcontrol BV	Dutch container fruit terminal
Kloosterboer BV	Dutch Fruit terminal
Capespan PLC	European importer of fruits
Caswell BV	Dutch IT service provider
VirtEx BV	Dutch IT service provider
Several growers and packhouses / cold stores	South African

In order to realize a practical workable structure within FRUITFUL project partners and other collaborating parties were grouped into three existing fruit supply chains. Each of these chains functioned as a FRUITFUL pilot supply chain with the three research institutes as the leading parties (one institute per pilot chain).

The basic structure of supply chains for refrigerated fruit from South Africa to The Netherlands as well as the 3 pilot chains is shown in Figure 1.

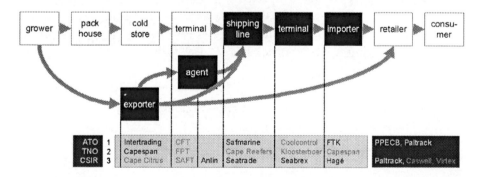

Figure 1. Supply chain of refrigerated fruit from South Africa to The Netherlands and structure of three pilot chains

Pilot chain 1 had a focus on transportation of mangoes and avocados in refrigerated containers (reefer containers) and was guided by A&F (ATO). Pilot chain 2 had a focus on shipping citrus and grapes with conventional reefer (refrigerated) vessels and was guided by TNO Inro. Pilot chain 3 had a focus on shipping citrus with conventional reefer vessels and was guided by CSIR Transportek.

The three pilot chains all belonged to the same logistical network. Most stakeholders in one of the individual pilot chains were role-players in other individual supply chains as well.

Through interviews and meetings with role-players in the pilot chains analysis of the pilot chains took place for processes, exchange of information (what, how, when, to whom) and for user requirements regarding improvement of the current situation. Subsequently the project team identified and specified the problem areas and prioritized them per pilot chain together with the industry. After that, solution alternatives were chosen for selected pilot items (as time and funding were limited only feasible items were selected). Solutions were further developed and finally tested in pilots.

Although the three pilot chains had different characteristics and each pilot focused on separate issues, together they covered the big part of the basic supply chain' or logistical network. The combined items of the three pilots are shown in Figure 2.

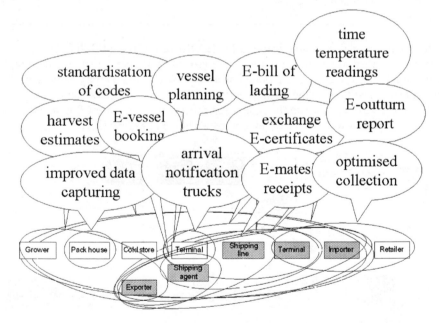

Figure 2. Combined pilot issues for the pilot supply chain

IMPORTANT RESULTS OF FRUITFUL

Combined pilot issues

- In order to implement electronic data exchange (EDI) in the whole supply chain all sending and receiving parties should use standard codes (e.g. port and country codes, commodity codes) and standardized message formats. During the project it became clear that very few standards existed although several initiatives were already running to correct this. In cooperation with the Fresh Produce Traceability Project (FPTP) and PPECB several standard formats and codes were introduced, including a standard format for exchange of quality information between the exporter and the importer.
- In one pilot chain all documents and exchange of information were collected during a physical pilot with 4 pallets of containerized fruit. This collection proved to be more than 40 items of documentation mostly consisting of faxes or hard copies. The collection showed the inefficiency and complexity regarding document exchange.
- In the pre-season planning, research showed that the harvest estimates and intakes at the depots that are used by exporters for all logistical planning are very inaccurate and may be logistically very inefficient and thus costly. A new strategy for planning vessels was developed for exporters, although the strategy was not tested during the FRUITFUL project. During the project an electronic web-based booking of containers was further developed and tested. Also a prototype web-based booking module was developed for booking pallets on conventional reefer vessels. The module was not tested.
- From the grower to the depot a couple of key issues were dealt with. Data capturing starts at the packhouse. For a solid and standardized backbone' that is required for EDI/integrated information systems, accurate and standardized data capturing at the packhouse is essential. This proved to be one of the bottlenecks and therefore standardization of codes, training of personnel and single pallet labels were considered to be requisites. Another bottleneck proved to be the Phytosanitary Certificate (PC) as a paper document. It is expected that European developments will make electronic exchange of PC possible within 5-10 years.
- Recommendations were made for improved notification of truck arrival at the port of loading in order to improve the logistical procedures in the port. These recommendations were not tested.
- Regarding preparing a load on a vessel and the voyage of the vessel possibilities were investigated for electronic Bills of Lading and an IT tool that facilitates putting a Bill of Lading on a closed website was tested. In addition the Mate's Receipt was made available electronically to various importers as well as to a receiving terminal in different ways.
- A website was further developed for the port of discharge under the FRUITFUL flag. The website was already under development by the discharge terminal. The data on the website related to information on availability and planning that is communicated between the terminal of loading, exporters, terminal of discharge,

importers and transportation companies in order to optimize the processes of the terminal of discharge. In addition, possibilities for electronic outturn reports were tested, including adding standardized information on quality at arrival.

Integrated electronic information system

During the FRUITFUL project it became clear that the climate was not suitable for an overarching centralized system and that the focus should be on interfacing existing facilities. In addition the information and communication (ICT) developments had developed rapidly during the past few years and were moving towards a preference for an ICT network structure. This outdated an overarching centralized information system.

It was concluded that bottlenecks for realizing fully integrated exchange of information did not lie on the technical side (hardware and software), but on the cost side and the human' side, e.g., education, procedures, data accuracy, mistrust, competition, institutional capacities, organization etc.

Development of level of collaboration within fruit supply chains

The South-African fruit export industry developed from a typical hierarchical structure before the deregulation in 1997 into a liberal market approach in 2002. It is a fundamental choice for each company to be either competitive in an open market or to collaborate in a network environment. In the case of the latter, a company would strive for connectivity, transparency and collaborative planning.

During the FRUITFUL project it became clear that the Capespan' pilot chain was much further developed towards collaborative planning than the other two pilot chains. This was not surprising as most partners in the Capespan' pilot chain were linked with one another in daily practice and worked together in a closed' supply chain as before the deregulation when *Capespan* was still *Unifruco* and *Outspan*.

One of the higher aims of the FRUITFUL project became working towards a network-oriented structure with transparent chains, connected information systems and collaborative planning. The project added to a much better understanding amongst the FRUITFUL role-players of each others' businesses and the identification of common aims.

More attention to producers and packhouses and smaller stakeholders

Data capturing starts at the packhouses (e.g. fruit specifications, pallet labels, etc.). Growers and packhouses were only involved in the FRUITFUL project through the exporters. This had not been a specific choice but was more due to how the project developed during the pre-phase.

As data capturing in the packhouse is the basis for exchange of information in the rest of the chain it was concluded that more attention should be paid to the start of the chain, not only in case of a FRUITFUL follow up, but in the whole fruit industry. In addition, in accordance with political and social developments more attention should be paid to the previously disadvantaged and/or emerging stakeholders.

A number of initiatives have been taken with the aim of empowering people from previously disadvantaged communities. The South-African Agri-Academy was formed (a non-profit organization) to facilitate and train members of these communities. Currently the Organised Fruit Industry is running workshops with the aim of developing a Fruit Industry Plan (FIP). This FIP is focused at determining the needs of these disadvantaged communities and drafting an action plan to address the necessary transformation. At the same time the Department of Agriculture has just (July 2004) published its document on Broad Based Black Economic Empowerment Framework for Agriculture'. This document aims to establish guiding principles for broad-based black economic empowerment in agriculture. It recognizes the challenges of globalization and the job threats facing farm and industry workers. Importantly it also addresses the forward and backward linkages within the total value chain within and between various commodities. The guidelines focus on various fields; these fields are: Agricultural Land, Human Resource Development, Employment Equity, Enterprise Ownership and Equity and Procurement and Contracts.

There is no preferential or easy entry into the fruit supply chain. Managing an efficient supply chain has certain requirements and these cannot be downscaled to suit new entrants from previously disadvantaged communities. The challenge is therefore to raise the level of skills and knowledge of these people to allow them to participate as equals. The question will be where to draw a line for producers / emerging stakeholders when it comes to having a chance to operate in the export fruit chain considering the level of education and technology that is needed for being able to be an equivalent business partner.

Contribution public-private partnership within FRUITFUL

The public-private partnership was crucial for the success of the FRUITFUL project. The presence of the three research institutes as independent parties removed distrust between competitive project partners and restraints could be bridged. During the project the partners started understanding more about each other's positions and problems and constraints and requirements as well as they started seeing more common goals.

The role of running and managing a complex project like FRUITFUL with so many partners in the North as well as in the South fits in much better into the expertise and core business of the institutes than the private companies. Such a role does not suit the companies. In addition the institutes are needed to keep a project running and to keep companies executing their task.

Through the public-private partnership much knowledge and insight was gained at two sides. It proved to be a requisite to have representative research partners in both South Africa and The Netherlands.

Desired institutional development for FRUITFUL aims

The South-African fruit industry was very clear on the role of the government in moving towards standardization. Important choices ought to be made by the government. In addition an independent body (e.g. fruit board) should keep the standards. This also had the effect that companies somehow were reserved in initiating changes (as long as it was not compulsory or no choices had been made from the governmental top).

Important lessons learnt

One of the lessons that were learnt was that within a project like FRUITFUL most time (of the research institutes) ought to be booked for bringing people (stakeholders) together, e.g., the facilitation of workshops and meetings and communication. Very often this is very hard to ŝell' when submitting a project proposal as most of the budget will then be spent on less tangible issues. Yet, one of the requisites for practical success and implementation will be that everyone understands the project aims, builds relationships and trust with all the other participants, and buys in to the project right from the beginning and after that stays involved and interested.

PRESENT SITUATION AND FOLLOW-UP

Presently (August 2004) the situation in the fruit supply chain from South Africa to The Netherlands has progressed further.
- The clock is ticking towards compulsory traceability due to the General Food Law (January 2005).
- The South-African fruit industry has become more receptive to collaboration and change with regard to information than was previously experienced.
- A growth can be identified of both the South-African and Dutch institutional development towards the fruit industry.
- A national Fruit Industry Plan for South Africa is being developed and will be implemented (CIAMD).
- Integration of previously disadvantaged and emerging stakeholders in the South-African fruit export industry has become a political priority.

A&F, TNO Inro, CSIR Transportek and the South-African – Dutch fruit industry have taken the initiative to develop a follow-up proposal. FRUITFUL 2 would be a follow-up of FRUITFUL according to the same successful formula of running a project with pilot chains. Main objectives of FRUITFUL 2 would be:
- Strengthening of the position of the fruit chains from South Africa to The Netherlands from a win-win point of view.
- Further developing integrated exchange of information in the fruit export chain from South Africa to The Netherlands, which will comply with international developments and requirements, in a practical way.

- Giving special attention to small stakeholders and previously disadvantaged partners according to the South-African policy and the policies of the project partners in The Netherlands.
- Initiating sustainable outcome of the project through appropriate transfer of knowledge.

The business partners of FRUITFUL 2 would (partly) not be the same as in FRUITFUL. In the follow-up the aims would join with the present climate and developments, and all the expertise that was gained through FRUITFUL would be applied.

CONCLUSIONS

The FRUITFUL project aimed to improve the exchange of information in the fruit export chains from South Africa to The Netherlands in order to improve logistical performance and quality of the fruit at the end of the chains. This was required to prevent any further loss of the position of South-African fruit on the world market. Improving the performance of this fruit export would be a win-win situation for the South-African and Dutch stakeholders:

- The project contributed to improved information exchange between chain partners. To name but a few examples, from a supply-chain perspective exporters were convinced to start using E-booking systems offered by shipping lines; the discharge terminal was given direct access to the South-African data system making the exchange of separate, non-uniform files of each exporter superfluous; the possibility was created to up-load outturn reports and storage information per grower immediately in the South-African data system. The project also accelerated the start of the development of a Fruit Information Plan.
- The project contributed to a better understanding between stakeholders of each other's position and problems. Employees who are familiar with other role-players in the chain only through telephone or e-mails concerning day-to-day business from their own perspective, were found to be able to solve some key issues simply by meeting these other role-players face-to-face to discuss common issues and clarify roles and responsibilities. Bringing operational people from business partners together also seems to be more effective than having only contacts on commercial level.
- The project was carried out in a(n institutional) climate that was not completely ready for implementation of the results.
- The formula of running the project through three existing pilot chains proved to be very successful.
- The public-private partnership as in FRUITFUL proved to be a very successful formula, although recommendations for further improvement were identified.
- Without a follow-up the benefits of FRUITFUL will go to waste.

WAY FORWARD

All stakeholders agreed that a follow-up would be necessary and beneficial and therefore a project idea has been submitted to the DGIS-LNV International Agro-Food Chain & Network Programme. All improvements made in logistics, quality, food safety, etc. are also necessary for emerging farmers, and this will get more attention. More research is needed to define the conditions that are needed for these (previously disadvantaged and emerging) stakeholders to be able to operate in the fresh fruit export chain.

REFERENCES

BuaNews, 2003. *Boost for black empowerment*. GCIS, Johannesburg.
Department of Agriculture, 2004. *Broad-based Black Economic Empowerment Framework for Agriculture*. Department of Agriculture South Africa, Pretoria. [http://www.nda.agric.za/docs/agribee/agriBEE.htm]
Fruit South Africa, 2004. *The face of fresh fruit in South Africa*. Available: [http://www.fruitsa.co.za].
McKenna, M., 2000. South African deciduous fruit industry under threat. *The Orchardist* (Dec.), 22-24.

CHAPTER 12

BRASCAN

How to capture value in the beef chain

MARCOS FAVA NEVES[#] AND ROBERTO FAVA SCARE[#]

[#] *Professor of Strategy and Marketing at the FEA-RP/USP School of Business and Economics, University of São Paulo, Brazil, and Coordinator of PENSA Brazilian Agri-business Programme*
E-mail: mfaneves@usp.br; www.fearp.usp.br/neves
[##] *PhD Candidate at the FEA/USP, School of Business and Economics, University of São Paulo, Brazil,*
and Researcher for PENSA Brazilian Agribusiness Program
E-mail: rfava@usp.br; www.pensa.org.br

Abstract. Brazil is rapidly becoming a major player in the world beef market. The Brascan Company is trying to capture more value in the beef business in order to enhance its investments. Opportunities are available for extending agro-food chains and networks that could contribute to this development. Two major strategies are discussed: (1) to become a large and reliable supplier in a beef chain, Brascan needs to grow horizontally; (2) to capture added value, Brascan needs to improve coordination in the vertical direction of the supply chain. The company is currently identifying strategies that could be helpful to achieve these aims.
Keywords: market segments; beef exports; quality control; vertical coordination

BUILDING AN INTERNATIONAL BEEF CHAIN: THE BRASCAN CASE

At the global level food moves largely from South to North, and Brazil is becoming an important supplier. In 2003, the country occupied the top position in world exports of soybean (38% of world market share), sugar (30%), beef (20%), coffee (29%), orange juice (82%) and tobacco (23%). Brazil was the second world exporter of soybean starch (34%), poultry (29%) and soybean oil (28%). In addition, Brazil delivers 16% of world pork exports, 4% of maize and 5% of cotton. The annual growth rate in these products has been 6.4% since 1990 (ICONE 2004).

The Midwest 'cerrado' region of Braz il is, according to Nobel Prize winner Norman Borlaug, one of the few remaining agricultural frontiers in the world, with almost 200 million hectares of new land suitable for agriculture.

R. Ruben, M. Slingerland and H. Nijhoff (eds.), Agro-food Chains and Networks for Development, 141-154.
© 2006 *Springer. Printed in the Netherlands.*

The Canadian-Brazilian group Brascan is involved in several activities. It is an asset management company, with a focus on real estate and power generation. The company has US$16 billion in direct investments and a further US$7 billion of assets under management. It owns 55 premier office properties and 45 power-generating plants. The investment policy is geared to areas where Brascan possesses a competitive advantage, and includes acquiring assets on a value basis with the goal of maximizing return on capital, building sustainable cash flows to provide certainty, reducing risk and lowering the costs of capital. Brascan recognizes that superior returns involve hard work and often coincide with different strategies.

The long-term goals of the company involve 12-15% annual growth in cash flow from operations, 20% cash return on common equity and 12% increase in the value of the company. In 2003 Brascan experienced a 31% increase in the company's registered value. Like other large multi-business companies, Brascan is increasing the return on capital, reducing risk by narrowing the areas in which the company operates, and broadening activities in these areas. This is the reason for entering beef production, where current operational margins are 5%.

The company faces a dilemma as to whether Brascan should invest or grow by broadening its activities or trying to capture more value on current activities. Opportunities have been identified in extending agro-food chains and networks to contribute to this development. Two major points choices emerge: to be a large and reliable supplier in a beef chain, Brascan needs to grow horizontally, but to capture value-added Brascan needs further vertical chain coordination. What should it do?

THE BEEF CHAIN IN BRAZIL: A GROWING BUSINESS

Brazil has shown spectacular growth in the beef business. Several factors have contributed to this growth: (a) internal factors of production including genetics, diversity, costs, availability of grasslands and technology; (b) external factors including the crisis of beef production in Europe (BSE), Argentina, Uruguay and recently, the USA.

The largest beef consumer markets are the USA, European Union, China and Brazil (see Table 1). In terms of import volumes, the largest markets are the USA, Russia and other former Soviet Union countries, EU and Japan.

Table 2 shows trends in per-capita consumption of beef. The largest consumers are Argentina, Uruguay, USA, Brazil and France. Data for Brazilian production from 1995 to 2003 are shown in Table 3. Strong growth took Brazil to the number-one position in 2003, when it exported 1.35 million tons and accounted for 23% of the international beef market (against 9% in 1999 – Table 4).

Table 1. Potential beef markets

Consumer market Consumption (thousand ton eq. c.)				Import market Imports (thousand ton eq. c.)			
	2002	**2003(*)**	**2004 (**)**		**2002**	**2003(*)**	**2004(**)**
USA	12,738	12,341	12,843	USA	1,460	1,363	1,510
EU	7,507	7,640	7,550	Japan	678	810	520
Brazil	6,437	6,273	6,400	Russia	660	590	650
China	5,830	6,116	6,484	EU	518	550	560
Russia	2,369	2,225	2,225	Mexico	489	370	250
Mexico	2,409	2,308	2,440	South Korea	430	444	200
Argentina	2,361	2,428	2,240	Canada	307	274	230
Other	10,300	9,152	9,183	Other	683	582	560
World total	49,951	48,800	48,544	World total	5,225	4,983	4,480

Note: (*) Preliminary; (**) Provisional   ; eq. c = equivalent carcass
Source: FAS/USDA (2004); 1 ton equivalent carcass (1 ton eq. c.) is standard measure in the data analysis, 1 kg of industrialized meat is equal to 2.5 kg of meat equivalent carcass and 1 kg of boned meat is equal to 1.3 kg of meat equivalent carcass.

Table 2. World beef consumption(kg/person/year)

Countries	*1997*	*1998*	*1999*	*2000*	*2001*	*2002**	*2003***
USA	43.1	43.7	44.1	43.9	43	44.2	41.9
Argentina	70	63.6	67.4	67.8	67.3	61.8	61.8
Brazil	39	38	36.3	35.8	35.6	35.8	36.2
Uruguay	66.6	72.2	71.3	61.2	51.2	60.2	56
France	25.9	26.6	26.9	27.4	23	25.8	25.6
Germany	14.7	15	15.2	15.5	13	14.7	14.7
Japan	11.5	11.7	11.7	12.	10.8	10.2	11
China	3.5	3.8	4.0	4.2	4.4	4.4	4.4

Note: (*) Preliminary; (**) Provisional
Source: Database selected from FNP Consulting – Anualpec (2003) based on USDA database. Elaborated by MAPA, October 2003

In 2004 world beef consumption was estimated at about 48,600 expressed in 1000 tons equivalent carcass. Beef consumption per country expressed in the same units for 2004 is estimated to be about 12,000 in the USA, 7,500 in the EU (including France and Germany), 6,500 in Brazil, 6,250 in China and only 2,400 in

Argentina. High consumption per capita is compensated by low population numbers in the case of Argentina, while low consumption per capita is compensated by high population numbers in China. Australia maintains a low consumption level of about 700.

Table 3. Brazilian livestock- sector evolution (1995-2003)

	1995	1996	1997	1998	1999	2000	2001	2002	2003*
Livestock (10^6 of animals)	154.1	152.8	153.2	155.6	157.3	160.8	164.1	167.4	177.6
Production (10^3 Ton Eq. C.)	6,768	6,794	6,406	6,491	6,539	6,583	6,892	7,143	7,530
Annual consumption per person	42.6	42.4	39.0	38.0	36.3	35.8	35.6	35.8	35.5
Exports (10^3 Tons.)	287	280	287	370	541	554	789	929	1,140

Note: (*) Preliminary
Source: FNP Consulting – Anualpec (2003) based on USDA database. Elaborated by MAPA, October 2003.

World beef production, estimated to be 50,047 (expressed in 1000 tons equivalent carcass) in 2004, more or less kept pace with world beef consumption (48,600 in 2004). Major producers were the USA with 11,700, Brazil with 7,800, the EU with 7,300, China with 6,300 and Australia with 1,900. Australia and Brazil have large net surpluses, which explains their position as leading exporting countries (Table 4).

Table 4. World beef exports (in thousands of tons)

	1999		2000		2001		2002		2003*		2004**	
	Vol	%	Vol	%	Vol	%	Vol	%	Vol	%	Vol	%
Australia	1,270	22%	1,338	23%	1,398	24%	1,365	21%	1,261	20%	1,300	22%
Brazil	541	9%	554	9%	789	14%	881	14%	1,175	18%	1,350	23%
USA	1,094	19%	1,119	19%	1,029	18%	1,110	17%	1,144	18%	195	3%
Canada	492	8%	523	9%	575	10%	610	10%	384	6%	420	7%
New Zealand	462	8%	505	9%	516	9%	503	8%	578	9%	560	9.5%
EU	949	16%	615	10%	546	9%	512	8%	400	6%	360	6%
India	222	4%	365	6%	370	6%	416	7%	465	7%	520	9%
Argentina	359	6%	357	6%	168	3%	348	5%	384	6%	420	7%
Others	508	9%	552	9%	419	7%	641	10%	638	10%	610	10%
World total	5,897	100%	5,928	100%	5,810	100%	6,386	100%	6,429	100%	5,880	100%

Note: (*) Estimate; (**) Preliminary
Source: Prepared by authors based on FNP Consultoria (2003) and USDA database

Brazilian exports in 2003 amounted US$ 1.5 billion, 30% higher than in 2002 (see Figure 1). In 2004, exports reached US$ 2.45 billion (63% higher than 2003). A quarter of Brazilian exports are directed at the European Union, whereas 55% of imports come from the EU. The Arabian markets are also becoming more important; although paying less, but there are no quotas, taxes are lower and there is market demand for different cuts than those in prevailing demand in the EU. This is very important for slaughterhouses, since more different products from each animal can be sold. In 2004 export efforts were progressing well and are expected to reach at least US$ 2 billion. There was a 30% growth in quantity and an almost 30% growth in value in the period January – June, compared with 2003. The largest growth in demand is reached from Russia, the United Kingdom, The Netherlands and Egypt.

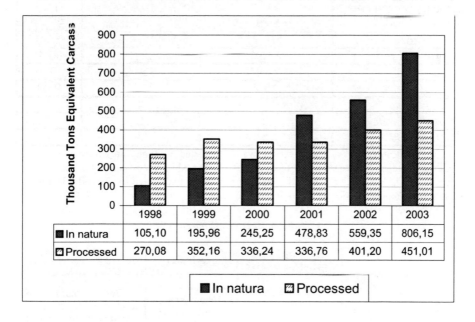

Figure 1. *Brazilian beef exports evolution by category*
Source: Prepared by authors based on FNP Consultoria (2003) and USDA database

The future prospects for Brazilian beef look very bright, despite the fact that, according to ICONE (2004), 61% of the total world beef market (US$ 25.3 billion) is closed to Brazilian exports. Four of the largest five importers do not buy from Brazil because of foot-and-mouth disease. Brazil is, however, a very strong participant in the open markets, which explains why the country is the world leader (see Table 4). Getting better access to closed markets would create new export opportunities, as would tariff reductions. Today, Brazil exports to more than 100 countries and Marfrig is one of the leading companies responsible for this market diversification.

For the sake of clarity we provide some figures regarding the competitiveness of Brazilian beef exports and the impact of tariffs and levies on beef prices. A Dutch importer (ANUALPEC 2003) pays US$ 3,700 for a ton of ¢ontra-filet' from Brazil (including US$ 200 for transport). Adding 12.8% import tax takes the price up to US$ 4,174. The full levy tariff, EU$ 3034/ton (US$ 3,731), also has to be added, making the final market price US$ 7,905 (e.g. 113% higher than the production

Table 5. Barriers to Brazilian exports

Product	European Union	United States	Japan
Sugar	160.8*	167.0*	154.3*
Alcohol	46.7*	47.5*	83.3
Milk	68.4*	49.1*	196.7*
Poultry (frozen cuts)	94.5*	16.9*	11.9
Swine (frozen)	50.6*	0.0	309.5*
Beef (frozen)	176.7*	26.4	50.0
Corn	84.9*	2.3*	95.4*
Tabacco	24.9*	350.0	0.0
Orange Juice	15.2	44.5*	21.4
Tariff Quota	7	4	1
Specific Tariff	8	6	4
Safeguards	5	3	2

Note: (*) Indicates that specific tariffs were converted in their *ad valorem* equivalent. Underlined indicates the existence of safeguards; shadowed indicates sanitary barriers.
Source: WTO, APEC, COMTRADE, USITC, TARIC – Elaborated by ICONE (Instituto de Estudos do Comércio e Negociações Internacionais)

Table 6. Beef: trade barriers

	TARIFF EXTRA - QUOTA	QUOTA	SANITARY BARRIERS
EUROPEAN UNION	12.8% + € 3034/t (98.2%) - refrig. 12.8% + € 3041/t (176.7%) - frozen	40,000 tons in Hilton Quota, 53,000 tons. in GATT Quota and 39,000 tons in ITQ Quota	Beef in natura: authorized for some regions of Mercosur considered foot-and-mouth disease-free
JAPAN	50.0% (s;s and half carcass refrigerated or frozen)	-	Demands a national territory free of foot-and-mouth disease
USA	26.4% Extraquota US$ 4,4 c/Kg 1.5% 'Intraquota'	696,000 tons.	Demands a national territory free of foot-and-mouth disease

Source: USITC, OMC. Elaborated by ICONE (Instituto de Estudos do Comércio e Negociações Internacionais).

price). On top of this, distributors' margins also have to be added (10-20%). More than 50% of Brazilian exports fall under this system. The rest falls under the quotas, and is subject to a tariff of 20% (Tables 5 and 6).

Several countries have quotas but do not use them due to the lack of production capacity. Licenses have to be paid for, however, to transfer a quota from one importer to another importer. Market information shows that there are still opportunities for growth in beef consumption in countries where per-capita consumption is not high (e,g. China only has an annual per-capita consumption of 4,4 kg of beef compared to 33 kg of pork). Prices for similar-quality beef are sometimes six times higher in The Netherlands than in the largest Brazilian supermarkets. What would happen to beef consumption in Europe if these high prices were reduced? Brazil is a cost leader, as it maintains extensive and natural methods of production as well as the production capacity that would enable it to keep up with the projected growth in consumption.

The long-term competitive position of Brazil in the world market can be analysed by considering four market segments where major beef volumes are sold and the particular supply position of Brazilian producers (see Figure 2 and Annex 1). We notice that major prospects for beef exports are available in the markets of the third and fourth quadrant where demand is growing and Brazil is still a minor supplier.

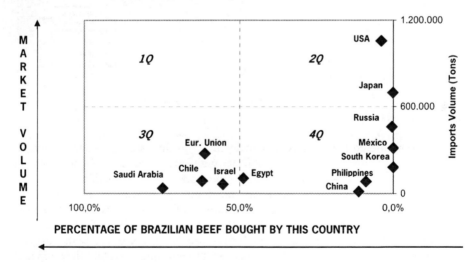

Figure 2. *Positioning of and Participation in Markets: Brazilian Beef Competitive Quadrants*
Source: Elaborated by authors based on USDA data

OPERATIONS OF BRASCAN IN BRAZIL

Brascan arrived in Brazil more than 100 years ago, and was one of the first multinationals to start operating in the country. It has several businesses, including agribusiness (cattle, land leasing for row and fruit crops, and timber), energy and other investments. It also has a bank and holds 40% of Accor in Brazil, as well as operating in hotels and employees' food coupons. Brascan strongly values long-term contracts and the company operates according to a network structure (see Figure 3).

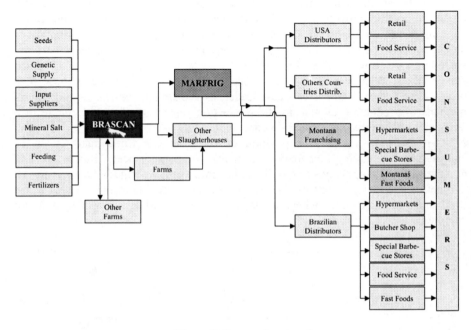

***Figure 3**. Brascan's network*

Source: Authors – Pensa 2004

Brascan entered the agribusiness sector in 1984, investing in a farm originally from the Swift group. Several investments were made and in 1999 the company started to act intensively, selling animals to the market through auctions.

Brascan now owns four farms in the southeast of Brazil: Bartira (14.400 hectares), Formosa (6,800 hectares), Mosquito (13,300 hectares) and Pirapitinga (17,900 hectares). Brascan has 40,000 cattle on these farms, 25% of which are used for artificial insemination. The main breeds are Nelore, crossed with Aberdeen Angus, Braunvieh, Brangus, Brahman and Santa Gertrudis. By 2007 they expect to have 50,000 cattle. It raises and sells weaned calves for further fattening. The breeding strategy focuses on including characteristics for precocity. All animals are traced and the brand name is Bartira', an Indian name.

Brascan has also diversified its operations. The farms also produce soybeans (100,000 bags/year), sugar cane (3,000 hectares), rubber (180,000 trees), pineapple (4,000 tons), American Quarter Horse' horses and pine and eucalyptus trees (for pulp and paper / furniture). The company also operates an environmental project, aimed at conserving the Mico Leão Cara Preta' monkey, which is threatened with extinction. There is competition for land in terms of returns per activity. The value of a hectare in São Paulo is US$2,000 and in Mato Grosso around US$100. Each Brascan company has an environmental task group, indicating the priority that Brascan attaches to this subject. In Cerrado they own 3,000 hectares of land that has been marked for nature conservation purposes.

BRASCAN BEEF NETWORK

Brascan excels in farm production. It has linkages with well-known research institutes in Brazil, including Unesp and Embrapa. Brascan supplies around 30,000 animals per year, either directly to slaughterhouses or selling calves to growers.

In *farm supplies*, the company has good relationships and dealings with major suppliers due to its size. In order to export more and to capture value, Brascan needs to improve its consistency and volumes, and also ensure it has well-organized supplies. To guarantee this, the company is investing in vertical integration of its genetic supply chain.

In terms of *production*, some new activities have been identified. Brascan is preparing itself for different types of long-term contracts in the beef chain. It perceives this kind of governance structure as an alternative for capturing value. In order to be able to comply with a medium contract with a distributor in the European Union the company would need 400,000 cows producing around 180,000 animals a year. At present it has 40,000 cows.

Brascan can sell calves or finished animals (ready for slaughter). The Brascan auctions are growing fast and are building up a good market reputation. They are selling calves with a 20% mark-up. Around 8000 animals are sold at six auctions each year. Brascan has been using Biorastro (linked to Eurepgap) to trace all its animals since 2001. It considers traceability as fundamental for selling in international markets. Brazil has introduced a governmental programme named SISBOV that collects data from private companies so that all animals in Brazil can be traced. It is not an easy task as Brazil is the biggest commercial producer of beef in the world and the country is very large, but a start has been made.

AN ALTERNATIVE TO VERTICAL GROWTH IN THE AGRICULTURAL CHAIN: THE MARFRIG EXPERIENCE

Now we come to the question how Brascan can capture more value in the beef chain. The market is growing, Brazil has become the leader and Brascan sees opportunities, one of which is with the Marfrig slaughterhouse and cooling company. Before examining Marfrig, however, it is important to discuss the processing part of the supply chain and in particular the functioning of slaughterhouses in Brazil. The major slaughterhouses were started by cattle producers. They vertically integrated production and processing and became food suppliers with some marketing functions. Most are still are family-managed groups. International companies were active in the beef industry, but their share has been reduced during the last 10-15 years (Table 7).

Table 7. Brazilian exports (US$ F.O.B.)

Company	JAN – DEC	
	2003	2002
Bertin Ltda.	432,585,717	357,719,233
Friboi Ltda	270,540,041	164,66,.904
Indústria e Comércio de Carnes Minerva Ltda.	148,225,985	92,946,846
Independência Alimentos Ltda	142,438,491	119,430,719
Marfrig Frigoríficos e Comércio de Alimentos Ltda	94,254,989	46,373,407

Source: Ministério do Desenvolvimento, Secretaria de Comércio Exterior.

The Marfrig Group includes four production centres and one distribution centre: Unit Bataguassú (MS), Unit Santo André (SP), Unit Promissão (SP), Unit Ribas do Rio Pardo (MS) and Unit Tangará da Serra (MT). The production units located in Bataguassú (MS) and Promissão (SP) have a slaughtering capacity of 1,900 heads of cattle per day, processing a total of 40,000 metric tons of beef per month. The distribution centre, located in Santo André (SP), has a storage capacity of 8,000 metric tons of food products.

Marfrig exports were valued at almost US$ 100 million in 2003, 100% more than in 2002. The company has introduced some innovative practices in the beef business, as its background is in beef distribution. The company's employees, about 2,000 people, received professional training, educational incentives, health care and nutritional assistance. A body of highly specialized technicians is in charge of the general supervision of all the processes and also takes care of the international protocols of quality and health.

The company's quality-control department makes use of a modern audit system, aligned with international standards of corporate management, to monitor all the processes of slaughtering, boning, packing, transportation and distribution. The quality programme implemented by the company involves a set of actions geared to the production of top-grade beef, meeting the rising demand from domestic and foreign markets. The first step of the programme is putting to pasture high-quality calves that will be slaughtered at roughly 30 months of age, accompanied by minimum standards for fat-covering, raising, livestock health and nutrition techniques.

The farms where livestock production occurs are no further than 300 km away from the slaughterhouses. This is a key factor for the production process, and guarantees better beef quality. The procedures recommended in this programme comply with international animal-welfare norms.

The quality-control programme also includes the certification of livestock origin. The process starts at the cattle ranges, where animals are identified with an earring and kept until they are slaughtered at the abattoir. All the animals currently slaughtered at Marfrig can be tracked trough the Bovine Identification System (SISBOV – *Sistema de Identificação Bovina*), guaranteed by Brazil's Ministry of Agriculture. The whole process is supervised and certified by the Biorastro system (visit to the cattle ranges) and HACCP (Hazard Analysis and Critical Control Points), both of which are internationally recognized by the FDA (US Food and Drug Administration), the European Union and the FAO/WHO Codex Alimentarius.

Another label of certification that the Marfrig Group has is granted by the Bio-Dynamic Institute (IBD – *Instituto Bio-Dinâmico*), which certifies the company's organic-meat process and is recognized by international entities. Marfrig undertakes several activities with meat distributors in Brazil, using the Montana brand in fast food, and the premium Montana Beef.

The domestic market for beef is very attractive due to high per-capita consumption. However, 50% of this market is informal, making it difficult for an organized company to operate. Moreover, the internal market absorbs meat and meat cuts that cannot be exported. With regard to competitiveness, it is expected that foreign companies will enter the Brazilian market. Once Brazilian production has access to other major beef-importing countries, Brazil will have to solve the tax problems and install a more rigorous tax-monitoring system. Several studies are being conducted and reforms are likely to take place in the next two to three years.

Brascan embarked upon a relationship with Marfrig in 2001. At that moment, Marfrig was just one possible marketing channel for selling animals, similar to a number of other slaughterhouses. After four years of building relationships with producers, a higher degree of trust has been established, some joint projects and studies have been started, and some joint planning takes place, but there are still no formal supply contracts. Brascan brings international buyers for Marfrig's products and has a good relationship with the company. However, this does not necessarily lead to better prices.

FURTHER PERSPECTIVES

Brascan needs to capture more value in the beef business in order to justify the investments in beef made by the head office. Opportunities to contribute to these developments are perceived in extending agro-food chains and networks. Two major alternatives emerged. Furst, to become a large and reliable supplier in a beef chain, Brascan needs to grow horizontally to a position where it maintains 400,000 cows. A key issue here is to identify which horizontal governance structures are appropriate for Brascan. Second, to capture added value, Brascan needs to increase vertical coordination throughout the supply chain. The company is currently identifying strategies to help it achieve this objective by tightening its relations with Marfrig.

ACKNOWLEDGEMENTS

The authors would like to acknowledge Renato Cavalini and Luis Fernando Della Togna from Brascan for their full-time support and Raquel Nascimento from PENSA, who participated in a first draft. This project is part of the European Union's SAFEACC project (www.globalfoodnetwork.org).

FURTHER READING

Berman, B., 1996. *Marketing Channels*. New York: Wiley.

Bridgewater, S. and Egon, C., 2002. *International Marketing Relationships*, Palgrave.

Camargo Neto, P., 2002. *Apresentação: Panorama Mundial da Carne*. Ministério da Agricultura, Abastecimento e Pecuária. Séc. de Produção e Comercialização.

Cavalcanti, M.R., 2002. *Relatório Beef Point – SIAL 2002*. Piracicaba.

Couglan, A., 2001. *Marketing Channels*. Prentice Hall, 6th ed.

Gemunden, H.G., 1997. *Relationships and Networks in International Markets*. Oxford: Pergamon.

Naude, P and Turnbull, P.W., 1998. *Network Dynamics in International Marketing* . Oxford: Pergamon.

Neves, M.F., Machado, C.P., Carvalho, D.T. and Castro, L.T., 2002. *Building Networks in the Global Beef Market* – The 18th Annual IMP Conference – ESC Dijon , 5 - 7 September 2002. p. 42.

Neves, M.F., Scare, R.F., Bombig, R.T. and Castro, L.T., 2002. *Choque de Marketing na Pecuária de Corte Brasileira* - Anais do 5° Congresso Brasileiro das Raças Zebuínas- Uberaba-MG; 20 - 23 October 2002.

Neves, M.F., Zylbersztajn, D., Machado Filho, C.P. and Bombig, R.T., 2002. *Marketing & Ações Coletivas em Redes de Empresas: O Caso da Carne Bovina no Mato Grosso do Sul* - Anais do XI Congresso da SOBER, Universidade de Passo Fundo (UPF), 2002, p. 181.

Neves, M.F., Machado, C.P., Carvalho, D.T. and Castro, L.T., 2002. *Marketing da Carne Bovina com Visão de Redes de Empresas ("Networks")* - Revista de Administração de Lavras. Vol. 4, n° 2, julho/dezembro de 2002, p. 73-85.

Rosembloon, B., 1999. *Marketing Channels* – 6th. ed. Chicago: The Dryden Press.

Stern, L., El Ansary, A. and Coughlan, A., 1996. *Marketing Channels*. New York: Prentice Hall.

FURTHER INFORMATION

ABIEC - Brazilian Association of Beef Exporter Industries www.abiec.com.br

ANUALPEC 2003 – *Brazilian Cattle Breeding Yearbook* – FNP Consultoria. São Paulo. 2003. www.fnp.com.br

BRASCAN Brasil - www.brascanbrasil.com.br

Brazilian Association of Nelore Cattle Breeders - www.nelore.org.br

Brazilian Beef Information Service - www.sic.org.br

Brazilian Institute of Geography and Statistics IBGE - www.ibge.gov.br

Brazilian Rural Society - http://www.srb.org.br

EMBRAPA Beef Cattle - http://www.cnpgc.embrapa.br

ICONE – Institute for International Trade Negotiations. www.iconebrasil.org.br

Marfrig Slaughterhouse Group - www.marfrig.com.br

Ministry of Agriculture Cattle Raising and Provision - http://www.agricultura.gov.br

Ministry of Development, Industry and Foreign Trade - www.desenvolvimento.gov.br

Montana Grill Group - www.montanagrill.com.br

PENSA - Brazilian Agribusiness Program – www.pensa.org.br

University of São Paulo - www.usp.br

USDA – United States Department of Agriculture. www.usda.gov

ANNEX A (background for Figure 2)

Table A1. *Big-volume markets where Brazil is not relevant*

Beef imports – metric tons per origin country				
USA	**2000**	**%**	**2001**	**Share**
Total imp.	**,**	**4%**	**1,056,004**	**(%)**
Australia	341,510	12%	383,719	36%
Canada	33,133	7%	353,671	33%
N. Zealand	212,379	0%	211,854	20%

Mexico	**2000**	**%**	**2001**	**Share**
Total imp.	**309,093**	**1%**	**312,996**	**(%)**
EU	239,681	0%	239,108	76%
Canada	42,046	41%	59,104	19%
Uruguay	13,765	-97%	450	0%

Russia	**2000**	**%**	**2001**	**Share**
Total imp.	**333,073**	**38%**	**459,756**	**(%)**
European U.	133,573	145%	327,582	71%
Ukraine	139,930	-35%	90,379	20%
Mongolia	15,249	-29%	10,765	2%

Japan	**2000**	**%**	**2001**	**Share**
Total imp.	**740,592**	**-6%**	**695,762**	**(%)**
EU	361,999	-10%	324,727	47%
Australia	332,617	-2%	326,453	47%
New Zealand	15,183	18%	17,954	3%

Source: USDA

Table A2. *Large-volume markets where Brazil is relevant*

Beef imports - metric tons per origin country				
European Union	**2000**	**VAR (%)**	**2001**	**Share**
Total	**299,185**	**-8%**	**274,869**	**(%)**
Brazil	150,239	12%	168,149	61%
Argentina	60,886	-56%	26,966	10%
Uruguay	20,906	-11%	18,571	7%

Source: USDA

Table A3. *Small-volume markets where Brazil is relevant*

Beef Imports- metric Tons Per Origin Country				
Egypt	**2000**	**%**	**2001**	**Share**
Total	**163,131**	**-36%**	**104,591**	**(%)**
EU	131,577	-97%	3,828	4%
Brazil	3,814	1237%	51,002	49%
India	24,230	51%	36,586	35%

Philippines	**2000**	**%**	**2001**	**Share**
Total	**88,224**	**-6%**	**82,710**	**(%)**
India	43,471	15%	49,815	55%
Australia	16,659	24%	20,641	23%
Brazil	1,892	279%	7,175	8%

Israel	**2000**	**%**	**2001**	**Share**
Total	**62,620**	**1%**	**63,055**	**(%)**
Brazil	13,469	159%	34,950	55%
Uruguay	28,088	-29%	20,020	32%
Argentina	16,521	-59%	6,741	11%

Chile	**2000**	**%**	**2001**	**Share**
Total	**88,873**	**-4%**	**85,328**	**(%)**
Brazil	29,067	82%	52,943	62%
Paraguay	21,366	20%	25,662	30%
Argentina	32,892	-87%	4,344	5%

Saudi Arabia	**2000**	**%**	**2001**	**Share**
Total	**30.234**	**18%**	**35,565**	**(%)**
Basil	3.139	745%	26,538	75%
EU	22.236	-100%	35	0%
Australia	848	550%	5,514	16%

Source: USDA

Table A4. *Potential markets*

Beef imports -Metric tons per origin country				
China	**2000**	**VAR (%)**	**2001**	**Share**
Total	**11,181**	**28%**	**14,269**	**(%)**
USA	6,606	3%	6,834	48%
Australia	2,398	12%	2,692	19%
Brazil	848	86%	1,576	11%

Source: USDA

CHAPTER 13

FISH IN KENYA

The Nile-perch chain

RONALD SCHUURHUIZEN, AAD VAN TILBURG AND EMMA KAMBEWA

Wageningen University, Department of Social Sciences
Marketing and Consumer Sciences Group
P.O. Box 8130, 6700 EW Wageningen, The Netherlands,
E-mail: aad.vantilburg@wur.nl

Abstract. With the introduction of large-scale fish-processing plants in Lake Victoria, Kenya, the structure of the actor network changed considerably. The domestic chain characterized by small-scale fishermen has become increasingly marginalized. Competition for fish between the domestic and export market is rather unequal and the drive to sell fish overseas has resulted in reduced local availability. The absence of banks and credit institutions increased the dependencies of fishermen on other parties in the supply chain. Improved financial systems to enlarge the ability of fishermen to acquire loans, government involvement in catch standards, and joint action for supply-chain governance system is needed to create a more competitive market and to provide the desired sustainability.

Keywords: small-scale fisheries; quality management; sustainability

INTRODUCTION

This case study describes the economic, social and environmental effects of the formation of a Nile-perch chain from the Lake Victoria region in Kenya to international markets after the 1970s. Initially, new technological developments in quality control, transport and large-scale processing boosted supply of fish through the catches and landings of Nile perch by local fishermen.

Foreign investments entered the Lake Victoria region and created a large processing capacity, but the benefits for the region appeared to be limited because of a loss of traditional jobs and limited added value for the local population. Small-scale enterprises dominated the upper part of the value chain from fishermen to the processing industry. This part of the chain was characterized by a lack of sufficient quality measures and control, oligopolistic or monopolistic power in markets and incompleteness of market information. The traditional domestic fish markets at the

R. Ruben, M. Slingerland and H. Nijhoff (eds.), Agro-food Chains and Networks for
Development, 155-164.

beaches and in the interior of the country have been marginalized due to the creation of a buying-agent system by the processing industry. The fishermen became dependent on loans from these buying agents to obtain access to boats and fishing gear, which in turn reduced their power in price negotiations.

The growth in the number of small-scale fishermen resulted in over-fishing – sometimes with illegal means – thereby making a sustainable future fish stock in Lake Victoria insecure. The international fish chain, as created by the processing industry, changed the fish-processing and marketing system considerably.

The case study discusses the impact of the international fish chain on the sustainability[1] of livelihoods of the local households of fishermen.

INSTITUTIONAL SETTING AND SOCIO-ECONOMIC IMPORTANCE OF THE NILE-PERCH CHAIN

The basic structure of the international supply chain and network of fish is presented in Figure 1. It affects the livelihoods of the local people because fishing was, and still is, a main source of income for households in the Lake Victoria area. About 80% of the fishermen in the Lake Victoria area in the mid-1970s depended on fishery as their primary source of income (Abila and Jansen 1997; Jansen 1997; Mitullah 2000).

Figure 1. *The Nile-perch chain from Lake Victoria, Kenya, to the customers in the EU and USA*

The Nile perch was introduced as a foreign hunter species into Lake Victoria in the early twentieth century. It proved to adapt very well to the lake's conditions. Access to the lake by fishermen was regulated and enforced by local authorities (Abila and Jansen 1997), as is often the case in communities with local commons or natural resources (Dasgupta 2002). This period was characterized by low investments in modern' equipment such as outboard engines. Before the institution of the international supply chain, the local fishermen traded the fish they caught mostly to women (the fishmongers) on the communal beach or in the village or town market.

This changed after 1979 (Figure 2) when new technological developments like refrigerated containers made it possible to transport fish over large distances. Rising international fish prices were very attractive[2] for fish originating from Lake Victoria (Gibbon 1997). The opportunity to earn foreign exchange by selling fish in the European Union or the United States led to an explosion of activities on the almost 300 landing beaches of the Kenyan part of the lake.

The decreasing number of landings in the 1990s (Figure 2) was caused by the raise of quality standards by the EU. The increase in landings in 1994 was based on a fundamental change of market conditions due to the opening of several big

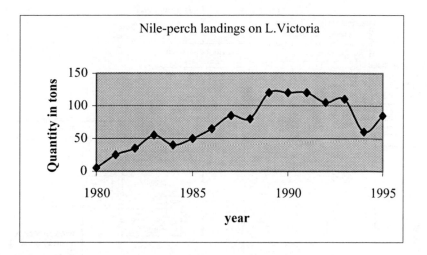

Figure 2: Nile-perch landings in tons during 1980-1995, Lake Victoria, Kenya (Abila and Jansen 1997)

processing plants (Gibbon 1997). The activities of these plants involved unloading, washing with chemicals, skinning, filleting, cleaning, packing and freezing to make the fish ready for the European market. A change in the number of landings could be caused by too many fishing boats or the fact that official regulations were disobeyed (LVFO 2004)[3].

The introduction of large-scale processing plants in the area changed the structure of the actor network considerably (Figure 3). This supply network connects the small-scale upper part of the chain with the large-scale processing industry and the international actors. The domestic chain is still characterized by small-scale fishermen and it has been marginalized by the export supply chain. Until recently, the development of the international supply chain for Nile perch was perceived to be a success story, providing foreign exchange to the area (ANF 2004). But some sustainability problems have arisen.

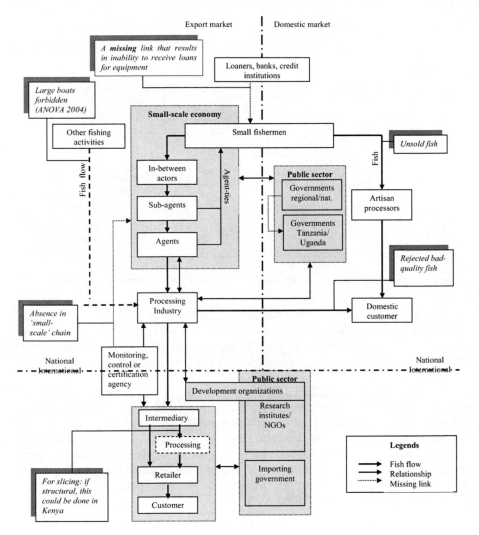

Figure 3: *The actor network of the international Nile-perch chain originating at Lake Victoria, Kenya*

MAIN BOTTLENECKS IN THE FISH CHAIN

This chapter discusses effects of the formation of the international fish chain on the sustainability of the livelihoods of the small-scale fishermen. Sustainability is assumed to have both an economic, social and environmental dimension.

Table 1: *Main economic, social and environmental dimensions of sustainability in the Nile-perch chain*

	Economic	Social	Environmental
Sustainability issues	**New market opportunities** -Limited regional benefits -Loss of traditional jobs -Limited added value for the poor **Upper part of the chain** -Insufficient quality measures -Product diversification at risk -Lack of monitoring, control and crediting **Market failures** -Overcapacity -No market transparency (info flow, monopsony)	**Livelihood issues** -Competition with domestic market; changing livelihood -Health and safety issues **Dependency and power distribution** -Ownership -Fisherman-agent structure **Trust and tradition** -Private optimization -Use of power -Specific circumstances	**The stock** -Tragedy of the commons -Long-term income effects **Loss of biodiversity** -Lack of enforcement -Damaging activities **Increased pollution** Water quality -Fossil transport fuels

Economic critical issues and sustainability

The formation of the international fish chain created new market opportunities that stimulated the local economy. More than 180,000 jobs in relation to fishery were created. As an example, the number of fishermen increased from 11,000 to 30,000 within two decades (Abila and Jansen 1997). However, consequences related to these new market opportunities were:

1) Regional benefits – investments in the local infrastructure or human-resource development – were low despite the growth in trade (O'Riordan 1996).
2) Traditional jobs like the – often female – fishmongers were lost (Abila and Jansen 1997). The working conditions for the factory workers tended to be insecure since they can be laid off within a few days.
3) The added value for the local population has been low. A large part of the value-added is taken by the buying agents, buying subagents and others occupying the chain between the fishermen and the processing industry. Also, the many people present in the chain for weighing, counting and carrying fish earn a part of the added value. The price paid for fish to the fishermen is sometimes about half of the price received by the buying agent that is delivering fish to the processing industry.

Secondly, there is not sufficient attention for several aspects of the small-scale fishermen's economy, although large fishing vessels are forbidden (ANOVA 2004).

Threats for the fishing community are:
1) Fish quality, as expected internationally, is guaranteed by the processing industry through adopting new quality guidance techniques. It is based on a strong selection of the quality of the fish that is purchased at the factory gate. There is insufficient quality control in this upstream part of the fish chain. Examples of the absence of quality monitoring in the upstream part of the channel are throwing of fish on the landing beaches; the insufficient use of ice on both the boats and the landing beaches, and long waiting times of the trucks at the beach competing for fish. This results in early deterioration of the fish.
2) The focus on only one species, the Nile perch, for export implies a considerable risk. About 50% of all the fish production at Lake Victoria is related to the Nile perch. Therefore, optimizing the level of risk in the chain through product diversification is required (ANF 2004).
3) There are almost no incentives for fishermen to change the current status quo.
    Several market failures can be observed in the fish chain. The early subsidies for processing factories by development banks and aid agencies led to underutilization of processing capacity in the area. In the downstream part of the fish channel, from the processing industry to the export markets, information on prices, quality, quantity and standards is quite clear. In the upstream part of the channel, between fishermen and processing industry, the fishermen do not have insight in the price that the processing industry is willing to pay for their fish, and this price can differ among beaches and buying agents. This is the result of an incomplete information flow (see Figure 4). The fishermen are dependent on the price that the buying agents are willing to pay. This is also due to interlocked markets, the obligation to sell fish to the buying agent who provides the loan for the boat or the fishing gear. Cooperation among buying agents – agreeing on which price to buy – is common, and leads to a kind of monopolistic buying behaviour on the beaches.

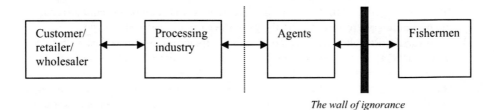

The wall of ignorance

Figure 4. The information flow from and to the actors, and the barriers to these flows

Social critical issues and sustainability

Fish is still an important part of the livelihood in fishermen villages. Before the development of the international fish chain the fishermen were fishing for local consumption. Nowadays, processed fish is mainly exported. Competition for fish between the domestic and export market is rather unequal due to the relatively high prices that are paid in the international markets and the related power of the fish-processing industry. The value of the export market in 2004 tended to be more than 90% of the total value of the landings although about 50% of the total fish landings have been used for the export market. Only the juveniles and low-quality fish, which are not accepted by the processing industries, are left for local consumption. To quote Abila and Jansen (1997): "the drive to sell fish overseas has resulted in very little of it being available locally". The nutritional security of the fishing communities and their domestic markets is threatened due to scarcity of fish because of the exports (LVFO 2004). Abila and Jansen (1997) indicated that 50% of the population at the lake does not get the minimum intake of calories that is required.

The low quality of the fish for the local market can result in health problems, but alternative sources of protein are scarce. Connected with the health issues are safety issues for fishermen at work. The small boats and limited rescue equipment make the job hazardous for the fishermen.

Second, the small fishermen are usually dependent on the buying agents and become victim of unequal power distribution in the chain:

1) There is a clear separation between local processors and the large-scale companies from the EU and US (ANOVA 2004). The investments in processing factories originated from domestic or foreign sources[4]. The international trade in fish products is primarily and most importantly in the hands of intermediaries who are not concerned with the upper part of the chain between fishermen and processing industry (Visgilde 2004).

2) Arrangements to supply fish by the processing industry make the fishermen dependent. The most common way is that companies or independent buying agents are in charge of collecting the fish from the landing beaches and bringing it to the processing factories, often through contracted subagents (Abila and Jansen 1997). The fishermen are subject to exploitation by these buying agents because they can make the fishermen dependent through granting loans without formal contracts for undefined periods. These loans are used by the fishermen to buy fishing equipment such as fishing nets or boats with outboard engines.

3) The agents buying fish from fishermen or middlemen have to carry the costs of fish that is not accepted at the gate of the processing industries.

The following factors can be considered to be barriers for development of the supply chain of fish:

1) Every actor is active in the chain, but each with a private objective to optimize his or her goals. A private actor will not change his strategy as long as there are no incentives for chain integration and development even if it would benefit the total fish chain without costing him/her anything (ANF 2004).

2) Several of the institutions involved in Lake Victoria fisheries are a kind of quasi-governments due to their close connections with the government and the power they hold. Regulations are difficult to enforce because of this distribution of power.

3) The fishermen and buying agents live in a society where a certain way of handling of fish, adapted to the local conditions, was common. These traditional values may become in conflict with recently introduced (or enforced) new

technologies and ideas. For instance, ice for cooling fish, although available, is still not widely used on the fishing vessels.

Environmental critical issues and sustainability

The first set of critical issues is related to the stock of Nile perch in the Lake. The tragedy of the commons (Varian 1999) is that the fish stock is common property but that every individual fisherman is trying to maximize his output. The resulting output is not in line with the public optimal level, which is based on the regeneration capacity of the fish stock. That this is already occurring is shown by a comparison of the daily catches of around 400-500 kg in 1981 and of about 100-150 kg in 1996 (O'Riordan 1996).

Secondly, there are other environmental issues connected to a loss of biodiversity:
1) Although the Nile perch is not native to the lake and actually is a hunter that was threatening the stocks of other fish (LEI 2004; ANF 2004), nowadays the largest negative impact on these stocks is the regular use of illegal fishing practices. For instance, nets are in use with mesh sizes that are too narrow for juveniles and other fish species to escape, although it is forbidden to catch fish smaller than 45.7 cm since March 2002 (ANOVA 2004). However, these nets are cheaper on the black markets.
2) The fish-breeding habitats are severely damaged due to the local fishing activities. This is mainly the consequence of the distribution of fishing rights in certain areas[5]: the main rights assigned to fishing areas are also the location of important breeding habitats. Over-fishing and damaging of habitats occur mainly in these areas (ANF 2004).

The government of Kenya did not sufficiently enforce the rules for protecting the natural resource base, mainly due to the fact that the focus was more on creating foreign exchange earners than on sustainability.

Thirdly, there is an increasing pollution of water (LVFO 2004) and air because of the use of boats, trucks and airplanes.

IMPLICATIONS FOR STAKEHOLDERS

In this section, implications for the stakeholders in the supply chain are briefly discussed.

The national actors: small-scale fishermen, processing industry, credit institutions and the public sector

The chain provided new opportunities for the region and has grown at a fast rate. The government has seen many sustainability problems arise and has made regulations to fight these problems, but lacked in the end the ability or willingness to enforce them. Especially the absence of banks and credit institutions increased the dependencies of fishermen on other parties in the supply chain. Involvement of banks and credit institutions, enlarging the ability of fishermen to acquire a loan, can lead to a more competitive market.

The international actors: trader, retailer and customer

An important motivation of the international traders not to be involved in the local fishery economy is because they feel that the local processing industries are more qualified to deal with this (ANOVA 2004). This results in a lack of knowledge of what happens at the domestic level characterized by lack of market transparency and traceability of each lot of fish (LEI 2004).

The international actors in Europe or the USA, although aware of problems within food chains, are not likely to be actively involved in changes in the supply chain. International traders are especially interested in the price, which needs to be competitive compared to other fish species (Visgilde 2004), the convenience aspect and the healthiness of the product (ANOVA 2004).

There seems to be a lack of cooperation between the national and international actors in the supply chain. Actors often look only after the part of the chain in which they operate to make a profit, which implies that new ideas or rules are difficult to implement. Change will not start from the local fishermen, neither from one single actor in the fish supply chain. A type of joint action or supply-chain governance system is needed to provide the desired outputs for both the channel participants and the consumers in Europe or the USA.

NOTES

[1] The term sustainability and the way this term can be viewed from an economic, social and environmental perspective is elaborated in this chapter.
[2] For the importers this was an opportunity to reduce the increase in fish prices (Gibbon 1997) given the relatively very low prices for Nile perch.
[3] Ugandan fisherman stated that the decline in fish landings was due to too many boats, nets or fishermen (for 33% of them), whereas 43% of them thought that widespread disobedience of official regulations is to blame (LVFO 2004).
[4] Of the twelve big processing factories, eight are in the hands of Asians, the others in the hands of westerners (Abila and Jansen 1997). These factories where in the past located in Nairobi, but all moved to the LakeVictoria region for faster processing and transport of the fish.
[5] Due to country regulations, some areas of the Lake are forbidden for fishery, whereas others are assigned to be legal to fishing.

INTERVIEWS

ANF, 2004. Interview with Dr. Paul V. Bartels, Division of quality in chains and food quality, Agrotechnology and Food Innovations Department, Wageningen University and Research Centre, 18th of June, 2004

ANOVA, 2004. Notes of Dr. Ruud Verkerk, Wageningen University, about a visit at ANOVA Den Bosch, 3rd April 2004, and an interview with Mr. Willem Huisman, Commercial director Anova Food Europe.

LEI, 2004. Interview with Ir. Jos G.P. Smit, fishery economist at the LEI (Agricultural Economics Research Institute), Wageningen University and Research Centre, 4th of June, 2004.

Visgilde, 2004. Interview with Mr. H.B.B. Floot, representative of Visgilde, a Dutch franchise organisation of fish retailers, 23th of June, 2004.

RELEVANT WEBSITES

ANOVA: www.anovafood.net
ICLARM: www.worldfishcentre.com
LVFO Lake Victoria Fishery Organisation: www.inweh.unu.edu

REFERENCES

Abila, R. and Jansen, E.G., 1997. *From local to global markets: the fish exporting and fishmeal industries of Lake Victoria: structure, strategies and socio-economic impacts in Kenya.* IUCN, Nairobi. Socio-economics of the Nile Perch Fishery on Lake Victoria Report no. 2.

Dasgupta, P., 2002. *Economic development, environmental degradation and the persistence of depriviation in poor countries*, Cambridge. [http://www.beijer.kva.se/publications/pdf-archive/Disc168.pdf]

Gibbon, P., 1997. *Of saviours and punks: the political economy of the Nile perch marketing chain in Tanzania.* Centre for Development Research, Copenhagen. CDR Working Papers no. 97.3.

Jansen, E.G., 1997. *Rich fisheries, poor fisherfolk: some preliminary observations about the effects of trade and aid in the Lake Victoria fisheries.* Centre for Development and the Environment, University of Oslo, Oslo. Socio-economics of the Nile Perch Fishery on Lake Victoria Report no. 1.

Mitullah, W., 2000. *Lake Victoria Nile perch fish clusters: institutions, politics and joint action.* Institute of Development Studies, University of Nairobi, Nairobi. IDS Working Paper no. 87. [http://www.ids.ac.uk/ids/bookshop/wp/wp87.pdf]

O'Riordan, B., 1996. *An assessment of the current status of Kenya's Lake Victoria Fisheries: a report for the Intermediate Technology Development Group.* IT Kenya, Nairobi.

Varian, H.R., 1999. *Intermediate microeconomics: a modern approach.* 5th edn. Norton & Co., London.

CHAPTER 14

ORGANIC CACAO CHAIN FOR DEVELOPMENT

The case of the Talamanca Small-Farmers Association

MAJA SLINGERLAND[#] AND ENRIQUE DÍAZ GONZALEZ[##]

[#] *Sustainable development and system innovation group, Wageningen University,*
Lawickse Allee 11, 6701 AN Wageningen, The Netherlands. E-mail:
maja.slingerland@wur.nl
[##] *Finished MSc studies at Wageningen University with thesis on ecotourism in*
Costa Rica, presently independent consultant in Colombia. E-mail:
zogiden@yahoo.com

Abstract. In de Talamanca region in Costa Rica cocoa production was abandoned in the late 1970s when yields dropped to zero due to Monilia. In the early 1990s, the Talamanca Small-Farmers Association (APPTA) gained success in promoting its revival. By creating contacts with buyers of organic cacao in the United States, APPTA was able to certify a significant area of cacao and to start exporting to the USA. Organic cacao had positive effects on farmers' income and on the environment. Recently a number of problems occurred: the USA buyer has withdrawn, the price of conventional cacao increased and the amount of organic cacao has increased as well, making it very costly to pay a premium for organic cacao. Costa Rica and Panama both produce only small volumes of organic cacao. Some cocoa producers started production of organic banana. APPTA has to cope with market instability due to excess supply and with low local production due to biological threats. APPTA has to find a reliable buyer and either to increase the volume of cacao or to pool its production with the production realized by COCABO in Panama. This latter option will lead to a number of associated challenges such as harmonization of quality, organizing logistics, tracking and tracing, elaborating contracts between cooperatives etc. Developing new commercialization channels towards Europe reveals three options: APPTA sells beans directly to a Dutch buyer; APPTA sells beans to a Costa Rican processor who in turn exports semi-manufactured products to Dutch producers; or beans are processed by Costa Rican industry and APPTA commits to selling semi-manufactured products to Dutch producers. Each of these three options has different repercussions on labelling, quality, environmental impact and profits. Additional options to improve livelihood of cacao farmers consist of diversification of the production at farm level and developing alternative sources of income, such as eco-tourism. The real challenge facing APPTA is to determine how to bring stakeholders with different interests and competencies together in an effective way to improve chain performance and to enhance farmers' livelihoods.
Keywords: organic cacao; marketing; chain performance; market instability; low productivity

R. Ruben, M. Slingerland and H. Nijhoff (eds.), Agro-food Chains and Networks for
Development, 165-177.

INTRODUCTION

The production of cacao in Costa Rica, particularly in the region of Talamanca, has gone through different periods. Cacao was produced and commercialized in this region during colonial times. Intertribal wars and colonial uprisings, however, hindered long-term agricultural settlements. It was not until 1860 with the emergence of banana production that agricultural development started (Somarriba and Beer 2003). Cacao came soon afterwards to replace banana plantations wiped out by *Fusarium oxysporum*. Cacao reached its peak in the 1920s and became the most important crop between 1940 and 1970 (Somarriba and Beer 2003). Cocoa prices soared in 1977-1978, but Monilia (*Moniliophthora roreri*) appeared in the late 1970s and production dropped to nearly zero (Somarriba and Beer 2003), depriving farmers of their only source of cash income. As a result, farmers abandoned their cacao plantations and shifted to subsistence crops such as maize, beans, rice and guinea grass. By the early 1990s, they lived mainly from these subsistence crops and poultry, selling only a very low proportion of their production on the market. Others kept their cacao plantations, although maintenance was poor, and combined them with trees such as avocado, citrus, cedar and laurel.

In the early 1990s, after quite a long interruption of cacao production in the region, the Talamanca Small-Farmers Association (APPTA) gained success in promoting its revival. With the support of ANAI, a non-governmental organization of US origin, APPTA reorganized the commercial aspects of cacao and established the conditions required for promoting and implementing organic production. By creating contacts with buyers of organic cacao in the United States, APPTA was able to certify a significant area of cacao and to start exporting to the USA. After this initial success, APPTA also made efforts to sell other products (especially banana) grown by its members under the rainforest and often mixed with cacao, but which were also used for consumption by the family. As a result of these efforts, APPTA obtained organic certification for the bananas produced by its members.

Recently, however, APPTA and its producers are being confronted with new challenges related with the optimization of the cacao chain in response to world market developments. An important issue is the complexity of selling organic cacao. A recent considerable increase in the price of conventional cacao in the global market is making it difficult to obtain a premium for organic cacao. Likewise, the supply of organic cacao has increased worldwide, putting even more pressure on prices. As the former buyer of the association has withdrawn, APPTA is now faced with the task of finding a new reliable buyer that will pay the premium and buy the total amount that is organically produced. In addition, the currently produced volume is low, mainly due to poor plantation management and continuous phytosanitary problems (fungal diseases). Processing possibilities in Costa Rica are limited and only a few comply with certification requirements. Substantial improvements are needed in these fields, especially as new buyers have both quantity and quality demands. APPTA also has to deal with the costs of certification.

The way these new challenges are solved depends to a large extent on the capacity of APPTA to readapt itself to this new context.

ORGANIC CACAO CHAIN

During the phases of formation, organization and implementation of the organic cacao chain in Talamanca, the outstanding player has been APPTA, which has promoted, fostered and supported the whole process. The history of organic cacao will be discussed, focusing on two aspects: first, a description of the cocoa production chain with information about actors and volumes; and second, a description of its impacts on farmers' livelihood and on the environment.

Cacao production in Talamanca

Producers of organic cacao and banana in Costa Rica are mainly located in Talamanca County, which is part of the Province of Limón in the south-eastern part of the country. The county is characterized by a tropical climate with an average rainfall of 4,000 mm and an average temperature of 25.6°C. It includes lands between 40 and 1,500 m above sea level in two main well-defined areas: the highlands and the valley. The highlands account for about 82% of the total area and 20% of the population, while the valley accounts for 18% of the area and 80% of the population. Hilly areas have substantial constraints for agriculture, with low fertility and high risk of erosion. In contrast, the valley receives less rain, has slopes of less than 13%, and moderately fertile soils, although these are being at high risk from flooding. This area, not surprisingly, has been more intensively used for agriculture, with basic grains, cacao, guinea grass and fruits being the most important crops (Damiani 2001).

In term of social composition it is important to mention that Talamanca houses two of the largest Indigenous Reserves in Costa Rica. Originally established as a single reserve in 1977 with a total population of 6,500 inhabitants, the separate Indigenous Reserves of Bribri and Cabécar were established in 1982. The Indigenous Reserve of Talamanca (IRT) Bribri covers an area of 43,690 ha, while IRT Cabécar possesses 22,729 ha. Together they account for 23% (664 km²) of the area and 45% of the population of Talamanca County. They are both part of La Amistad National Park and the Talamanca-Caribbean biological corridor (Damiani 2001).

Farmers in Talamanca grow cacao as part of a production system that includes shade trees and rainforest. Most of the organic-cocoa producers in Talamanca are smallholder farmers who usually grow a mix of crops cultivated under the rainforest. After the diseases of the 1970s farmers abandoned their cacao plantations; many slashed and burned the areas with cacao and started to grow subsistence crops such as maize, beans and rice. For cash income they depended on timber, fishing and hunting. Others maintained their cacao plantation and combined cacao with trees such avocado, citrus, cedar and laurel. This production system required little work and farmers still harvested and sold some cacao at the end of each year. No labour was invested in phytosanitary or soil fertility measures.

APPTA (Talamanca Small-Farmers Association)

Costa Rican farmers and farmers' associations started organic cultivation of crops in several places at the same time in the mid-1980s, in response to crop diseases, high costs of pesticides and health problems caused by chemical inputs. The initiatives were unrelated but had in common that farmers started to experiment with organic fertilizers and pesticides.

ANAI, an NGO of US origin that started working in Talamanca in the mid-1980s, promoted reforestation activities among indigenous communities. Later on, its role was crucial in the creation of APPTA. ANAI encouraged farmers to create an organization that could serve collective interests such as collective marketing of products and attract foreign donors interested in the implementation of projects that involved the preservation of the rich environment of Talamanca. The Asociación de Pequeños Productores de Talamanca (APPTA) was created in 1987 and has since played a key role in the growth of organic agriculture in the region and in the access of small farmers to organic markets.

The first collective task that APPTA undertook was building and opening an input supply store, which is still operating. In addition, APPTA worked with environmental organizations and NGOs to promote rainforest conservation. Soon afterwards, it received support from the Inter-American Foundation, enabling it to strengthen the association by constructing buildings and purchasing equipment.

While APPTA was quite successful in attracting international funds for the conservation of the rainforest in a region with indigenous communities, by the late 1980s several members were arguing that it needed to change its focus towards more sustainable activities. These discussions marked the beginning of a more active role in the search for possible markets for their products, which eventually led to contacts with buyers of organic products. APPTA played a key role in achieving access to organic markets in three major ways:

o Identifying the possibility of organic certification for the dominant production systems among small farmers in Talamanca. In fact, organic production in Talamanca did not involve a substantial change, in contrast to production in other regions. This was because, when looking for market opportunities, APPTA made contact through ANAI with buyers of cacao in the United States who were looking for regions in developing countries where cacao plantations had been abandoned for several years due to pests and diseases, and who were promoting the idea of obtaining organic certification for these plantations.

o Organizing the marketing of organic cacao and banana. Due to the high cost of transactions involved in negotiations with single farmers, the presence of APPTA was instrumental for the organic-cacao buyers, as it was able to organize an efficient marketing system, purchasing the product from farmers and delivering to buyers in a timely, economic and convenient manner.

o Setting up and managing a monitoring system to ensure that all farmers use organic technologies. This has been the most important task carried out by APPTA. The association was able to organize an efficient system that is decentralized and based on members' participation. In fact, instead of organizing a central team of technicians who permanently visit farmers, as is done in many farmers' associations elsewhere, APPTA created 'local committees' in the different villages. These committees have worked very well because other members of the community recognize their roles and their decisions are fully respected.

So far we have considered the role of APPTA as promoter of organic cacao production in the region. The results of its other activities are even more impressive. By 2000, over 1,000 members of APPTA had obtained organic-producer certification for more than 2,000 hectares of cacao and banana (Damiani 2001). In the same year APPTA exported 210 tons of organic cacao, of which 160 tons went to the USA and 50 tons to Europe. By 2001 the export of organic cacao through APPTA had increased to 300 tons. In the same year about 2,300 ha was certified by APPTA, and cacao – either with or without banana – was grown on this area under rainforest (Damiani 2001). In 2003 there were already 1,170 organic producers in 50 communities in the Talamanca region. In the same year cacao farming was practised on 3,016 ha, with an average farm size of 10 ha (3.6-134 ha). Cacao plantations within these farms were on average 1.3 ha. The average cacao plantation is now 21 years old (1-80 years), with an average annual yield of between 100 and 200 kg/ha. In good years the total cacao yield of APPTA farmers is about 300 tons, although in bad ones it can drop to as low as 40 tons, as was the case in 2001. To date one third of all Costa Rican cacao comes from Talamanca.

Other relevant actors

There are other important stakeholders (Andrade and Detlefsen 2003) besides APPTA in the cacao chain. Although most of them are not exclusively related to Talamanca and do not solely trade in organic cacao, their role within the production and commercialization process of products from Talamanca is essential.

The cacao that is processed in Costa Rican cacao chains is not only produced within the country. An important portion comes from abroad, in particular from Panamá, where small cacao producers are organized in the Cooperativa Cacaoterra Bocatorena (COCABO). Conventionally produced beans are collected by COCABO and undergo processing into liquor blocks by the Costa Rican Cacao Products Company (CCP) or by FINMAC, a Dutch cacao farmer. The FINMAC liquor blocks are processed by a local company, Gallileto, into chocolate for Costa Rican consumers. This chain is based on conventional cacao because Costa Rican consumers are not willing to pay the premium for organic cacao. The liquor blocks from CCP are sold to the USA in the conventional cacao chain. Organically produced beans from Panamá also go through COCABO to CCP in order to be processed into liquor blocks. These blocks used to be sold by Organic Commodity Products Inc. (OCP, which recently withdrew from the chain) in the US market for a premium. Organically produced beans from the Talamanca region in Costa Rica are collected by APPTA and formerly followed one of four routes: 1. direct sale of beans to European buyers; 2. direct sale of beans to OCP (in USA and further sale to USA buyers); 3. processing at CCP into liquor blocks and sale through OCP to US buyers; 4. processing at CCP into liquor blocks and direct sale to European buyers.

Since the withdrawal of OCP from the chain, two of the more common alternatives no longer exist. Figure 1 summarizes the different actors in the chain and their relations.

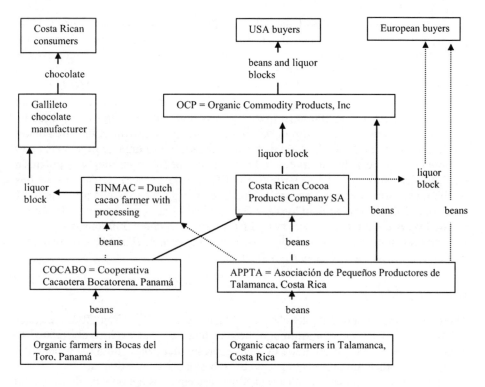

Figure 1. *Cacao chain*

Certification, grades and standards

For a country to be able to export organic products to the EU a number of minimum standards regarding 'organic' certification must be complied with. Until 1990, certification firms were based in Europe and the USA, making certification a costly affair. While laws and institutions allowing for locally based certification of firms were supposed to reduce the certification costs for farmers, trust relations with foreign certification firms dominated, obliging local firms to make partnerships with foreign ones, increasing costs again. Until 2001, Eco-Lógica was the only registered certifying firm in Costa Rica. Currently there are several additional national certification firms: Aimcopop, OCIA (US), BCS Öko Garantie (Germany), Ecocert (France), and Skal (The Netherlands).

  Two types of labels are important at present: organic (O) and fair trade (FLO). Farmers from APPTA or COCABO regard both organic and FLO labels as valid.

Problems tend to arise within the chain when processing takes place, is done by FINMAC. The Dutch cacao farmers are not entitled to the FLO label as they are not a local smallholder themselves. Figure 2 depicts how certification flows through the commercialization chain.

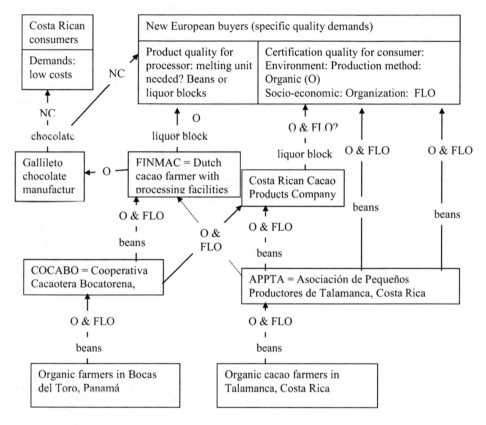

Figure 2. Diagram of organic (O), Fair-trade (FLO) and Non-certified cocoa (NC)

Effect of organic production on farmers' incomes and quality of life

Organic production in Talamanca has had significant positive effects on farmers' incomes and quality of life (Triana 2003). With the emergence and growth of organic production, farmers were able to start selling cacao again, and they obtained prices significantly higher than in the conventional market. This is important because Talamanca is one of the poorest regions in Costa Rica. In 2000, for instance, APPTA paid its members an average of 1 USD per kg of organic cacao against 0.40 USD per kg for conventional cacao. With the introduction of organic banana income became diversified. The combination was appropriate because, while cacao has two harvest peaks a year, banana can be harvested every two or three weeks all year round. Cacao and banana together provide 32% of farmers' revenues with another 37% coming from forest products that are part of the organic cacao and banana production system (Damiani 2001). If only cash income is considered, cacao and banana represent 62% of the income. Organic cacao and banana have also proven to be a feasible alternative for preventing health risks related to conventional agriculture.

Nevertheless, although APPTA is FLO-registered, farmers still suffer from poverty as they generally have a low cacao yield (average of 150 kg per ha) and hardly any other source of income. It is certainly worthwhile exploring additional options such as ecotourism or the promotion of products from other components of the system. The transformation of cacao waste into a more valuable product would be a convenient task to pursue.

Effect of organic production on the environment

The effect of organic production on the environment (Triana 2003) has been very positive. It has helped to safeguard a highly diverse system. More than 90% of Costa Rican floral diversity is found in Talamanca, and so are 1000 of the 1300 fern species found in the whole country. Organic cacao and banana have contributed to protecting this system because plantations are characterized by a thick groundcover that prevents erosion and leaching, and they help to recycle residues within the system. Plantations also host an important number of plant and animal species (see Table 1). The area of organic cacao serves as a buffer zone between agricultural area and natural forest and therefore contributes to the conservation of the biodiversity in the Talamanca-Caribbean Biological Corridor (Martínez et al. 2003).

***Table 1**. Plant and animal species in plantations*

System	Tree species	Mammals	Birds	Birds on CITES list
Natural forest	85	51	130	44
Shaded cacao	35	25	130	34
Shaded banana	14	9		

Sources: Giracocha (2000); Reitsma et al. (2001)

NEW CHALLENGES IN THE CACAO CHAIN

So far the organic cacao chain has gone through several steps: its formation, development, organization and implementation. It has even become properly embedded within a policy and legal framework. Training, research and extension supporting the chain are also in place. Nonetheless, world market conditions are changing continuously and all actors in any chain need to respond to these changes quickly to be able to maintain their presence and their results. To a large extent adjustments are achieved by optimizing the chain, yet remaining flexible.

Like many other actors in the chain, APPTA is now facing new and varied challenges regarding organic cacao. The present decade has brought new developments. In general there are two categories of challenges: market instability both in volume and in price due to excess supply, and low local production levels due to biological threats and poor management, which ultimately worsens quality standards.

Market instability

Volume and price are the two crucial factors endowing market instability (Hinojosa et al. 2003). On the world organic-cacao market many new producers have emerged, of which the relatively new presence of the Dominican Republic is notable, due to the large volume of cacao it produces. Not surprisingly, this event has led to a considerable increase in supply and a respective drop in prices. To solve this issue the buyers may consider focusing their efforts on increasing consumers' demand. In the case of APPTA, the challenge in this respect is greater due to the disappearance of a key chain actor, OCP Inc., a former organic-cacao buyer. The association needs now to ensure its relations with existing buyers and that prices are guaranteed even under current conditions. At the same time, it needs to find a new reliable buyer of organic cacao that will accept even larger volumes in a market that is rapidly becoming saturated.

The organic-cacao chain is also under pressure due to the increase in the price of conventional cacao. Political instability in Ivory Coast, the major producer of cacao in the world, has led to extremely low volumes of cacao during the last two years, resulting inevitably in booming prices. APPTA is now in a position of having to convince its organic-cacao producers to keep on producing organically even though the prices of conventionally produced cacao are now as high as the prices commanded by organic cacao. This situation is also encouraging farmers to focus on other crops like organic banana, making cacao only a second priority, and some are simply abandoning it. From an environmental point of view this is likely to be undesirable. From a farmers' income point of view it may be attractive in the short term to concentrate on banana alone. However, a diverse system with both cacao and banana is less vulnerable to market fluctuations.

At present Costa Rica has ample laws and institutions in place that give it preference with respect to the export of organic products to Europe, and therefore it is likely to have a relative competitive advantage compared with other providers. An increase in volume and improved performance in the chain can make APPTA even more attractive for big buyers.

Low production and low farmers' income

Besides price and volume fluctuations, low yields are the most important problem facing cacao farmers in the region. In Talamanca, production of organic cacao is extremely low per farm, per hectare and per tree. There are three main reasons. First, the number of cacao trees per ha is low and therefore the number of trees per farm is small. Second, the age of the plantations is relatively high, and they are therefore no longer so productive. Lastly, most of the plantations are not well maintained, which facilitates the presence of pests and diseases, especially when the cacao trees are old and little fertilization takes place. APPTA needs to help farmers increase their cacao production, but without endangering the certification or the biodiversity at farm level.

DEVELOPMENT IMPLICATIONS AND FUTURE PERSPECTIVES

As described above, the challenges facing APPTA are considerable. Despite these circumstances developments are beginning to happen that should help to overcome current restrictions. APPTA is definitely not alone in its struggle. Support is present in the form of existing and proposed projects with other actors and stakeholders. Currently, for instance, new actors are entering the scene, trying to replace OCP Inc. and its USA-oriented chain with a chain towards The Netherlands.

CREM, a Dutch consultancy agency, has done some studies on organic cacao and on the Costa Rican situation. They are currently assisting APPTA in trying to find an appropriate and interested Dutch cacao buyer for Costa Rican organic cacao. The particular arrangement chosen will depend on many factors. Nevertheless the following three options have already been identified as possible ways forward (CREM 2002a; 2002b): first, APPTA sells beans directly to a Dutch buyer; second, APPTA sells beans to a Costa Rican processor who in turn exports semi-manufactured products to Dutch producers; and third, beans are processed by Costa Rican industry and APPTA commits to selling semi-manufactured products to Dutch producers. Each of these three options has different repercussions on labelling, quality, environmental impact and profits.

The latter options in no way exclude other alternatives. APPTA for instance is taking up other organic products and exploring other marketing strategies so that it can become involved in new product–market combinations. At the same time, APPTA's equivalent in neighbouring Panama, COCABO, represents an important opportunity for joining forces to improve the commercialization of their products.

New opportunities are also on the horizon with regard to the problems of low productivity and pest diseases. CAB International is an international NGO active in Costa Rica and Panama. Together with CATIE (Tropical Agriculture Research and Higher Education Centre), it is carrying out projects on biological control of cacao diseases, on rational pesticide use in cacao, and on genetic improvement. A combination of cultural practices and biological control with an appropriate mixture of antagonists (mycoparasites) can significantly reduce losses due to pests. Likewise the project 'Organic Cacao and Biodiversity', managed by CATIE, has played a major role in the development of organic cacao production and commercialization in Talamanca. The objective has been to enhance sustainable production and biodiversity conservation of at least 300 indigenous farms producing organic cacao in Talamanca, favouring biodiversity conservation in the Mesoamerican Biological Corridor (Somarriba and Harvey 2003). The project has 4 components: on-farm biodiversity   conservation;   sustainable   production   and   commercialization;

reinforcement of local organizations and on-farm biodiversity monitoring. It would be most welcome if this project could be continued for the coming years.

Efforts of other actors have been aimed at the promotion of new economic alternatives such as ecotourism (Díaz González 2004; CREM 2003). The idea behind such initiatives is to make the organic cacao system more viable by generating additional income from sources other than cacao.

Finally, Wageningen University is joining forces with CATIE and CABI to work at integrated supply-chain development, including options for quality improvement and monitoring, reduction in costs of certification, improved timeliness and volumes of delivery, appropriate benefit-sharing among stakeholders, etc. Additional options include transforming cacao waste into new products to generate additional income and exploring options with regard to the production of organic chocolate products within Costa Rica. For the latter option, organic ingredients other than cacao (e.g. milk) need to be available. Producing these ingredients might offer opportunities to develop new organic chains in other sectors.

Another path is the development of specific breeding objectives concerning disease resistance, at the same time taking into account consumers' and processors' wishes regarding quality. The working methods are fully participatory, including all stakeholders in meetings, training and capacity building of both individuals and institutions, and working with farmers' field schools to keep responsibilities as much as possible at the farmers' level.

CONCLUSIONS

The role of APPTA in the development of organic cacao in Talamanca has been outstanding. Its actions have not only helped cacao production to re-emerge, but also to widen the economic alternatives of Talamanca's local farmers. Organic production of cacao and banana has had a great positive impact on the incomes and quality of life of small farmers in Talamanca because both crops are important and complementary sources of cash income. Perhaps many inhabitants still live under adverse conditions, but they certainly have a platform from which to move forward.

The association has been successful in providing new opportunities for its members through certification of organic cacao and banana production and the establishment of new commercialization channels. APPTA has created economies of scale by managing higher volumes with lower transactions costs. It has also been successful in creating business relations with foreign firms and in creating a monitoring system that effectively ensures that all members comply with organic methods of production, one of the key requirements of the organic certification process for small-farmers' associations. APPTA provides intensive training to its members and in this way has attracted and maintained the interest of all its members.

Besides economic advantages, the role of APPTA as a promoter of organic production systems has also provided positive effects on the environment of Talamanca, which is one of the most diverse ecosystems in Costa Rica and at the same time one of the regions most affected by the expansion of commercial agriculture in rainforest areas. Organic cacao and banana systems have contributed to the conservation of the rainforest and wildlife.

Despite all the positive benefits mentioned so far, as a result of the recent circumstances referred to in the second half of this article, APPTA is being forced to reinvent its own role to overcome the current challenges imposed by the market. The association has to find a reliable buyer able to replace OPC Inc. Likewise, it urgently needs to improve plantation yields and product quality, especially due to the emergence of new producers that probably have higher investments and better incentives. The presence of COCABO is certainly crucial in pursuing a bigger and better commercialization channel. A partnership between these two farmers' organizations is a goal that needs to be achieved in the short term for the good of both associations.

APPTA is definitely not alone in its struggle, however. It is receiving support through existing and proposed projects. Various actors (European and local) are interested in finding solutions to all current constraints. The real challenge facing APPTA is to determine how to bring stakeholders with different interests and competences together in an effective way to improve chain performance and to enhance farmers' livelihoods.

ACKNOWLEDGEMENTS

We would like to thank Mr Andre Frezac-Coloneth from COCABO-Panama for his presence at the conference. He gave his reactions to the paper and highlighted the COCABO side. Thanks also to Ulrike Krauss from CABI for her regular comments on the issue and for facilitating contact with Mr Frezac-Coloneth.

REFERENCES

Andrade, H. and Detlefsen, G., 2003. Principales actores de Talamanca. *Agroforestería en las Américas*, 10 (37/38), 6-11.
CREM, 2002a. *Jungle chocolate from Costa Rica?* Available: [http://www.crem.nl/pagesen/nbwinter2002.html] (February 2004).
CREM, 2002b. *Opportunities for sustainable cocoa chain between Costa Rica and the Netherlands*. Available: [http://www.crem.nl/pagesen/handel.html] (February 2004).
CREM, 2003. *Rainforest Alliance and CREM Partnership*. Available: [http://www.crem.nl/pagesen/nbwinter2003.html] (February 2004).
Damiani, O., 2001. *Organic agriculture in Costa Rica: the case of cacao and banana production in Talamanca*. International Fund for Agricultural Development, Rome.
Díaz González, E., 2004. *Ecotourism as a means for community-based sustainable development: La Cangreja National Park case study, Costa Rica*. MSc-thesis Wageningen University, Environmental Policy Group
Giracocha, G., 2000. *Conservación de la biodiversidad en los sistemas agroforestales cacaoteros y bananeros de Talamanca, Costa Rica*. CATIE, Turrialba. Tesis de Maestría, CATIE
Hinojosa, V., Stoian, D. and Somarriba, E., 2003. Trade volumes and price tendencies in international markets of organic cocoa (Theobroma cacao) and banana (Musa AAA). *Agroforesteria en las Americas*, 10 (37/38), 63-68.
Martínez, J., Somarriba, E. and Villalobos, M., 2003. Cacao orgánico y biodiversidad. *Agroforestería en las Américas*, 10 (37/38), 4-5.
Reitsma, R., Parrish, J.D. and McLarney, W., 2001. The role of cacao plantations in maintaining forest avian diversity in southeastern Costa Rica. *Agroforestry Systems*, 53 (2), 185-193.

Somarriba, E. and Beer, J., 2003. Diagnosis of agroforestry in indigenous Bribri and Cabecar small organic cacao farms in Talamanca, Costa Rica. *Agroforesteria en las Americas,* 10 (37/38), 24-30.

Somarriba, E. and Harvey, C.A., 2003. How to integrate sustainable production and conservation of biodiversity in indigenous organic cocoa plantations? *Agroforesteria en las Americas,* 10 (37/38), 12-17.

Triana, A.M., 2003. *Sustainability in the cocoa chain: identification of the relevant aspects, stakeholders perspectives and potential sustainability indicators for the cocoa chain.* MSc thesis Wageningen University, Environmental Policy Group

CHAPTER 15

THE NOVELLA PROJECT

Developing a sustainable supply chain for Allanblackia oil

LAWRENCE ATTIPOE, ANNETTE VAN ANDEL AND SAMUEL KOFI NYAME

SNV Netherlands Development Organisation, P. O. Box 30284, Airport, Accra, Ghana

Abstract. The Novella Project, a collaboration between a commercial company (Unilever), an international Non-Governmental Organization (NGO) (SNV Netherlands Development Organisation), local NGOs, local businesses, collectors, transporters and processors, aims at developing a strong, effective and sustainable international supply chain for Allanblackia oil. Success of the chain will increase when more local farmers and collectors will find the additional incomes attractive enough to get involved. A strong chain, with clear business opportunities, will encourage the local communities not only to protect the Allanblackia trees, but also to offer similar protection to all trees in the forest, thereby contributing to explicit sustainable forest management and maintenance of biodiversity. Unilever has already shown commitment to developing a sustainable supply chain by offering to support local communities to plant more Allanblackia trees on their farms and closer to the communities. The project's choice to focus on the empowerment of women is strategic as their participation can lead to improved livelihoods and stability of family incomes, which are important objectives of the project. For the commercial partners, including Unilever, a strong market chain is expected to translate into an acceptable return on investment. The chain is in its development stage and is confronted with a number of challenges. The collection of seeds from wild-growing trees in dense tropical forest and low tree density in village areas lead to high transaction costs for collection of the seeds along the chain. Also being a new market chain intense information, education and communication is required over a long period throughout the project areas to transfer knowledge, skills and project information to current and potential supply-chain actors. In the complex partnership different interests risk to compete with common defined objectives. There is the challenge of increasing involvement of local businesses and investments in the key areas of the chain. SNV has a crucial responsibility in addressing these challenges by supporting different partners involved in the project.

Keywords: partnership; economics; development; biodiversity; institution building; learning; non-timber forest product; sustainable forest management

INTRODUCTION

Establishing strong market chains and networks for products from developing countries is one important way to contribute to economic and social development,

179

R. Ruben, M. Slingerland and H. Nijhoff (eds.), Agro-food Chains and Networks for Development, 179-189.

through employment, increased incomes and participation of poor people in value-added economic activities. In rural areas of Africa the majority of the people are engaged in agriculture, including the collection of wild- growing non-timber forest products. Even though much time and effort of the local economy is spent on these activities the values created are still very low due, in part, to lack of access to important markets and the low quality of outputs. Developing effective agro-food chains can ensure increased incomes through markets and value-added activities. The more value is added to various links of agro-food chains the more likely it is that some of it will fall in the hands of the primary producers, often located in rural areas of developing economies, many of them women. This is often a good signal for others to participate in activities related to the market chain. Increased employment will also, hopefully, lead to more incomes. Typically the stronger and more established the market chain and networks, the more interesting it becomes for local entrepreneurs to play an active role through increased investment of their time and other resources in order to reap the benefits of added value.

THE PROJECT AND ITS OBJECTIVES

The Novella Project is a collaborative project between Unilever, SNV Ghana and local non-governmental organizations (NGOs) to develop a strong, effective and sustainable supply chain for Allanblackia oil. It is a unique partnership between a multinational business, local and international NGOs and businesses with diverse interests in a collective pursuit of a common agenda. Allanblackia (locally called Sonkyi) has been identified as an edible fat that can be used to produce food products such as margarine. Prior to the Novella Project local populations collected the seeds for personal use as they had no significant commercial value. The fat produced was used for human consumption and for the local production of soaps. Communities of local farmers collect the Allanblackia seeds for sale aiming at increasing their income. In Ghana, the Allanblackia trees grow wild in the tropical rainforest belt, mostly in the cocoa-growing areas of the Central and Western Regions. The population of the Western and Central Region is 1,924,577 and 1,593,823, respectively, with 27% and 48% of them living below the poverty line (Ghana Statistical Service 2002) and with the majority clustered around the poverty line. Most of the people of these regions are farmers living in rural communities. Collection and sale of Allanblackia seeds may be an opportunity to increase their income.

A unique initiative

Unilever initiated the Novella Project to develop a new supply chain for Allanblackia oil. Ghana was identified by Unilever as a pilot country for the Novella Project because of the abundant presence of the Allanblackia tree in several of its regions, its political stability, the presence of local partners and some initial trials on the extraction of oil with Ghanaian oil processors.

Unilever sought, at an early stage in the project, to pursue a development agenda in addition to the clear commercial agenda. Unilever therefore approached other partners who could bring this company and collectors (gatherers) together to start a pilot project, which was named the Novella Project because it is a novelty in many respects. This is the first time Unilever is engaged in this kind of joint public-private partnership to develop a supply chain based on both commercial and development agendas. This is also the first time a concerted effort is being made to find a commercial value for the Allanblackia seed.

Key project objectives

At the inception of the project all partners agreed on the following common objectives:
- to contribute to the enhancement of rural livelihoods
- to contribute to the sustainable management of forest areas and reserves in Ghana, as well as maintenance of biodiversity
- to encourage the participation of women in all aspects of the Allanblackia supply chain
- to contribute actively to the non-use of child labour (ILO convention).

Economic importance and development agenda

For Unilever the clear benefit would be a strong, effective, sustainable and profitable supply chain. The project is also expected to create business opportunities for local entrepreneurs to participate in all the processes that lead to the production of the final products in Unilever's factories. Local small businesses will be active in the field of buying, storing and transporting the seeds and extracting the oil from the seeds for Unilever. More value will thus be added to the product before it is exported. But this has to be done in a sustainable way, including a strong emphasis on biodiversity conservation and sustainable forest management and non-use of child labour. In the end additional employment and incomes will be expected to improve rural livelihoods. Participation of women in all aspects of the project will ensure equitable distribution of the wealth created. Within the framework of the Novella project, SNV Ghana facilitates the implementation of the development agenda through the local NGOs.

KEY PROJECT PARTNERS, PARTNERSHIP ROLES AND GUIDING PRINCIPLES

The key partners in the Novella Project have been Unilever, SNV Netherlands Development Organisation, Friends of the Nation (FoN) and Institute of Cultural Affairs (ICA) (the latter two being local NGOs with a lot of community knowledge and experience). Other institutions involved with promoting the Allanblackia business include IUCN – The World Conservation Union, Swiss State Secretariat for Economic Affairs (Seco), Forest Research Institute of Ghana and Technoserve.

Unilever

Unilever's role is essentially that of project manager with a particular focus on the supply chain and its actors. In the pilot phases Unilever provided resources to the local NGOs to provide information, education and communication in the project communities. They also provided the overall framework for the Novella Project including strategy and planning. They set the targets for the year, deploy resources for the purchase of the seeds and resolve specific supply-chain issues with the support of the other partners. Currently Unilever has recruited field staff to assist with some of the activities the NGOs performed in the pilot phase. Unilever guarantees to buy specified quantities of the oil fat at a pre-set price from the processors. This enables the collectors to focus on collecting seeds instead of looking for buyers. Further they guarantee fair prices for all participants in the supply chain, which motivates the collectors, buyers, transporters and processors to feel adequately compensated for effort and resources invested in the market chain.

FoN and ICA (local NGOs)

The local NGOs have focused on propagating the message of the Allanblackia business including the need to pursue all the elements of the development agenda. They encourage local communities to collect the seeds and to keep them for the buying company, which is required to follow the trail of the local NGOs, to buy the seeds during the harvesting season. In the process the NGOs help interested communities to elect focal persons, ensuring during the process that democratic principles are followed and also that women are given a fair chance to present themselves for election as focal persons. The NGOs have done a fair amount of informing, educating and encouraging the communities to ensure that those who participate in the collection at the local level understand the business best practices of the Novella project and are prepared to play by the rules. The NGOs also monitor strict adherence of the collectors to the tenets of the development agenda.

Additionally the NGO field staff, working hand in hand with Unilever field staff, endeavour, through their activities, to empower the local communities to play an active role in determining the nature and direction of the Allanblackia business, including their ability to negotiate fair and reasonable prices when the supply chain is clearly established. They explain the elements of the pre-set price to the local communities. As part of activities in the second phase of the pilot the local NGOs, with the support of Unilever and SNV, facilitated the formation of collectors and the focal persons into local Allanblackia groups to ensure adequate focus on the development of the supply chain. It is expected that the groups will have the opportunity, through regular interaction, to discuss pertinent issues regarding their participation in the supply chain.

SNV Ghana

SNV's role has continued to be to give impetus to the development agenda and to provide important market linkages. SNV has, since the inception of the project, been assisting in enhancing the capacities of the local NGOs to be able to perform their roles as frontline agencies that carry the business message of Allanblackia to the local communities. SNV assists both Unilever and the local NGOs to design appropriate messages as well as develop capacity of the NGOs and Unilever field staff on appropriate modes of delivery for maximum effect. SNV also provides training on the integration of gender principles into the Novella Project and also conducts monitoring visits to ensure strict adherence to the development agenda.

Guiding principles

It became clear from the outset that there would always be a critical need for very clear guidelines and principles, acceptable to all partners, to enable a group of such diversity of interests to work together towards a common goal. These finally emerged through open and frank discussions among the partners. The essential elements are as stated below:
1. All parties acknowledge and respect the actual and potential role each organization can play in the Novella Project.
2. All parties respect each other's independence and choices considering the means and instruments selected for the implementation of the development agenda.
3. All parties shall strive for maximum transparency.
4. All parties shall strive for fair pricing structures.
5. The partnership is founded on principles of equity and all parties recognize elements of responsibility and accountability that would engender such equitable partnership.
6. All parties shall strive for a timely, adequate and fair flow of information.
7. The collaboration contributes to environmental sustainability and good governance as a means to making a contribution to poverty reduction.
8. The collaboration shall enable enough room for other interested parties to be enlisted, from time to time, to assist and facilitate the successful implementation of the Novella Project.
9. Adherence to local and international law including respect for ownership rights.
    Admittedly, it has not been easy for all the partners to abide by all of these principles. Nevertheless, through dialogue and regular evaluation, the partnership is still thriving.

KEY ACTORS IN THE SUPPLY CHAIN

The key actors in the supply chain of Allanblackia seed oil in Ghana include the collectors, the buying company, the transporters, the processors and Unilever, which buys the processed oil for export. These are supported by a number of NGOs and development organizations, whose role for the moment remains critical to activate the Allanblackia business until it becomes commercially viable. These NGOs are not part of the supply chain and therefore do not collect, buy, transport or process the seeds.

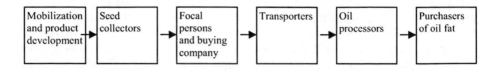

The Collectors

It was originally thought that the Allanblackia seeds could be found in sufficient quantities on local farms and on the fringes of the rainforest. It was also thought that most of the collectors would be farmers, who will see collection of the seeds as additional income-generating opportunities. Two phases of the pilot have taught us that although the collectors are predominantly farmers (farming being the main activity in the communities) it requires much additional effort, including walking long distances, to find the trees and the seeds in sufficient quantities to become an attractive commercial activity. The collectors are expected to lodge their collections with the focal persons, or at least stay in contact with the focal person to know when the buyers will come. Monitoring activities by both SNV and the local NGOs have found sufficient evidence that women play key roles in collecting seeds, while no evidence was found that children participate in the chain.

Focal persons

The focal persons currently mobilize the collectors within their communities and encourage them to collect the seeds during the harvesting season in return for a commission on every kilo of seeds sold. At the moment they form an important link between the buyers and the collectors. A clear advantage is that they reside in the local communities, while the buyers mostly come from the cities. As the business develops, it is expected that the focal persons and local businesses may form their own buying networks to buy and sell directly to the processors or to the buyers from the cities.

Buying company and transporters

Unilever has contracted a buying company to purchase the seeds on its behalf and to transport them to the processing plant near Accra. The buying company deploys its own resources to buy the seeds and is only required to sell the seeds to Unilever at a pre-set price. The contractual arrangement between Unilever and the buying company does not, however, preclude others from buying and selling the seeds. In fact almost all the actors will welcome more buyers to ensure timely purchase of all the seeds collected in the communities. The buying company liaises with the NGOs and Unilever field staff to determine the areas where seeds might be readily available to purchase. In the actual phase the terrain is vast and the quantities of collected seeds in the communities are small; therefore the buying company is unable to deploy resources to all the communities to purchase all the seeds. This discourages those whose seeds were not purchased from collecting more.

Oil processors

Currently a private company processes the seeds for Unilever. It has previously invested in infrastructure to process other oils for Unilever. It receives the seeds from the buying company for processing. In the third harvesting season the processor is also involved in purchasing the seeds directly from the collectors.

Purchase of oil fat (Unilever)

Unilever guarantees the processor a vast market for the Allanblackia seed oil at an agreed price. Through this process Unilever takes steps to ensure that all the safety and quality standards are met. This is an important incentive for actors in the supply chain.

ACHIEVEMENTS AND LESSONS LEARNT

Halfway through the third harvesting season and the pilot phases there are certainly some achievements and lessons to share. Yet many challenges remain.

Achievements

Several achievements can be mentioned at this stage. A solid partnership between public and private entities has been established with a clear commitment to develop the supply chain on sound business principles. There is a high level of awareness about the Novella Project in many communities where the seeds can be found. Various community entry and education formats have been tested. There is increased awareness about sustainable use of natural resources in the project communities. Some farmers regret ever cutting the many Allanblackia trees they thought were useless. Some farmers have agreed to plant Allanblackia trees and there is commitment to invest in propagation and domestication of the Allanblackia seeds. Participation of women in the project has been encouraging.

In the domain of knowledge and skills acceptable standards for handling and processing of Allanblackia seeds has been established (Amanor et al. 2003) and various researches have been carried out and others initiated to understand the business of Allanblackia (Attipoe et al. 2004). Actual partners show commitment to expand the scale of the project in Ghana and Tanzania and to commence in Nigeria, Cameroon and the Congo basin. The project raised interest of other development partners including the UNDP, Novib, GTZ, Seco, Intercooperation and Technoserve. Their involvement will, as expected, expand the scope and coverage of the supply chain.

Lessons learnt

The partnership has so far been a very rewarding experience for all the partners. It is clear that private and public organizations can work together to achieve common goals and objectives once the objectives, roles and responsibilities are clear. To achieve this clarity it is important to do a critical assessment of the mission, vision and resource bases of all partners. There should be a clear definition of roles and responsibilities among partners including who should coordinate the activities of the partnership. This should not be taken for granted. Furthermore, open and constant communication among partners is critical for the partnership. If possible the formats and standards for information sharing and communication should be discussed and agreed, including the levels and types of information that partners should have access to. The development agenda costs money and requires careful preparations to pursue. It has to be sufficiently clear to the partners, at the inception stage, how the project will be financed. This was not very clear in the Novella Project.

Another important lesson is learning a creative way to enhance access to important markets for the poor. In this case Unilever has guaranteed the market with the additional responsibility of assuring product safety and quality. The farmers in their individual capacities would not have had easy access to the market. Not only will their numbers be too low, but also the cost of investing in safety standards would be prohibitive at this stage. Involvement of local businesses and business interests is very critical to maintain local community interest in the project. Efforts should be made to include them.

Finally our experience has been that in this kind of partnership there are enormous opportunities to learn new things, new ways of doing things and new strategies, through open, frank and transparent engagements with partners.

CHALLENGES AND THE WAY FORWARD

To a new supply chain, developing from scratch for a product for which there was hardly any commercial value, there are bound to be challenges at different levels. The challenges are identified and activities are planned to meet them.

Partnership communication and coordination

The partnership is a novelty for all members. It is complex, involving multiple and sometimes contrasting objectives, orientations and methods, of partner organizations and stakeholders who are learning to adjust to each other's pace of work. There should be commitment to this learning process. It has been very difficult to communicate and coordinate activities among the partners. This meant that not all the partners had access to information in a timely manner. This is being addressed through a newly established Partnership Committee to be chaired by SNV.

Unilever had expected the NGOs to be fully committed to the achievement of the development agenda in the communities they were already working in. They were expected to use this experience to reach out to new communities. That commitment, it is assumed, should also challenge them to find additional resources to support the project. That is far from happening in the immediate future. The capacity of the NGOs to deliver has been weak in the two pilot years. The NGOs and service providers have created awareness through community entry, posters, public campaigns and interactions with the communities, yet most communities have not fully accepted the project. Education has to continue until the local communities realize the potential that is hidden in the Allanblackia seeds and can totally commit to its development. This has been an important learning point for the project. There has to be a realistic and fair assessment of the motivation and resource capacity of future participants in the project.

At the end of the evaluation after the second season, it became very clear that to achieve the targets the project has to expand to cover many communities scattered over the rainforest. The failure to achieve the set targets has a number of underlying causes: price, roles of and incentives for focal persons and buyers, volume and density of trees, motivation of local business.

Price

Feedback from stakeholders and field visits by all partners suggest that the going price of 1000 Ghanaian Cedis (the equivalent of 10 eurocents) per kilo is low and does not adequately motivate the collectors and focal persons. In some cases, however, with intensive education and persuasion some collectors and focal persons have renewed their commitment to the project believing that one day the chain will develop to the extent similar to cocoa. The strength of that commitment was greatly tested when we found that some of the collectors are not prepared, even where they have already collected the seeds, to sell them to the buyers. On the one hand there is a price beyond which the end product cannot enter the external market; on the other hand a price below which the collectors have absolutely no motivation to collect the seeds. This has been the real dilemma for the project and a challenge we accept with continuous education, information and communication.

All the partners recognized the price issue as a critical success factor of the project and have shown commitment to finding a solution. There have been very open and frank discussions among all the partners including a close scrutiny of Unilever's pricing formula. SNV has been conducting a price study to determine, from the point of view of the collectors, what are the main elements of a fair price for a product still in development.

Role of focal persons and buyers

Purchasing the seeds from the collectors is a great challenge at this moment. The volumes are still very low despite the huge potential that exists. There is no doubt that, just like cocoa and oil palm, once the project reaches critical mass businesses will find sufficient interest to warrant significant investment. At the moment, it seems, this has to be engineered in a very creative manner. The focal persons can play a very important role in the whole supply chain of the seeds. They have primary contacts with collectors and in most cases know them very well. Because the business is not sufficiently attractive at the moment the focal persons have not shown much commitment. The low volumes realized mean that they did not earn enough from commissions to be motivated. Also, not everyone in the communities accepted their role as buying agents. Some preferred to sell directly to the buying company. In some cases the focal persons did not have the appropriate equipment to weigh the seeds. Their role is being streamlined as we move into the next phase. SNV is developing an entrepreneurship development/awareness programme to sharpen the entrepreneurial competencies of the focal persons and generate enough business interest in the project within the communities.

Research has shown that there are indeed not enough seeds from wild-picking to make the project commercially viable for both Unilever and the communities. The partners are currently preparing a prospectus to raise 20 million euros for propagation and domestication of the Allanblackia trees to guarantee the quantities that will ensure commercial viability in the long term.

Motivation of local business

There is yet an important issue that has to be addressed to boost community interest in the project. Perhaps one of the greatest challenges we face with the Novella Project is how to motivate local businesses (businesses within the project communities) to participate actively in the supply chain of the Allanblackia seeds. In part this is because the potential business benefits are still not too clear to local entrepreneurs. Another reason is the lack of capital to invest in a venture that is still rather exploratory at this stage. Yet local business participation will perhaps be the best guarantee for sustainability of the project. Local entrepreneurs should be able to set up businesses around collecting, storage and transporting the product to the Unilever processing centres. The benefits accruing from local participation in increasingly longer segments of the chain will, hopefully, encourage more local people to participate in the collection of the seeds, which is the major weak link in the supply chain. Additionally local business participation is the only way we can put real incomes into the communities. The way it is currently structured most of the business and the benefits will only go to processors, buyers and transporters, who do not reside in the project areas. It is a real challenge to change this.

CONCLUSION

There is enough interest among the partners to continue with the project. All partners have renewed their commitment to the next phase of the project to ensure that the environment is not destroyed, that communities gain additional income through fair prices for their efforts, and also that women will continue to be the major beneficiaries of the project. The partners have all planned key activities for improving performance of all the actors in the supply chain as well as the supporting institutions with the view to supporting the development of a strong supply chain that provides good incentives for all stakeholders. Chaired by SNV a Partnership

Committee has been formed to improve communication and coordination among partners.

REFERENCES

Amanor, K., Ghansah, W., Hwathorne, W.D., et al., 2003. *Sustainable wild harvesting: best practices document for harvesting of Allanblackia seeds*. Project document.

Attipoe, L., Oppong, K. and Koranteng, F., 2004. *Novella price study*. Project document.

Ghana Statistical Service, 2002. *2000 Population and housing census: summary report of final results*. Ghana Statistical Service.

CHAPTER 16

DEVELOPING A SUSTAINABLE MEDICINAL-PLANT CHAIN IN INDIA

Linking people, markets and values

PETRA VAN DE KOP[#], GHAYUR ALAM[##] AND BART DE STEENHUIJSEN PITERS[#]

[#] *KIT Royal Tropical Institute, Amsterdam, The Netherlands. Tel. +31 20 5688711, Fax + 31(0)20 5688444. E-mail: p.v.d.kop@kit.nl*
[##] *Centre for Sustainable Development, Dehradun, India. Tel. +91 135 2733823, Fax +91 135 2733111. E-mail: alamcts@vsnl.com*

Abstract. In recent years the demand for medicinal and aromatic plants has grown rapidly because of accelerated local, national and international interest, the latter notably from Western pharmaceutical industry. At present, resource-poor people in India's poorest state Uttaranchal collect plants from the wild in order to complement their meagre incomes. Due to continued collection and increasing market demand, numerous plant species are threatened with extinction. For rational and regulated collection, strong local communities or strict governmental control measures are necessary. High risks, transaction costs and lack of trust among chain actors prevent smallholder producers from taking up cultivation of medicinal plants. Public–private collaboration is suggested as a way of reducing these constraints and to secure market access to small producers. Such collaboration can provide a promising mechanism for establishing the conditions for the establishment of supply chains in the initial stages of development.
Keywords: indigenous species; biodiversity protection; public–private cooperation

INTRODUCTION

The medicinal plant sector in Uttaranchal, a Himalayan state in northern India, can provide an important source of income to the rural population, especially because returns from traditional crops are declining (Alam 2003). Uttaranchal's unique climate, its locally available expertise, motivated farmers and NGOs and a supportive government policy provide a strong base from which to take advantage of the growing national and international demand for medicinal plants (Belt et al. 2003; Alam and Belt 2004). The main advantage of medicinal plants for small producers lies in the fact that, compared to bulky and perishable commodities, they have a

R. Ruben, M. Slingerland and H. Nijhoff (eds.), Agro-food Chains and Networks for Development, 191-202.

higher value per unit volume. This makes them particularly attractive for remote, mountainous areas with transport limitations.

In this paper, we analyse the opportunities for, and constraints on, developing medicinal-plant chains in Uttaranchal. The paper specifically aims to identify the role of medicinal-plant chains in poverty reduction; the basic conditions for successful integration of small producers in the medicinal-plant chain; and the institutional infrastructure required to support a pro-poor medicinal-plant chain. The paper is based on action research conducted by KIT Royal Tropical Institute, the Institute for Applied Manpower Research (IAMR) and the Centre for Sustainable Development (CSD). The study involved fieldwork and multi-stakeholder consultations to discuss research findings and identify pathways to the development of a pro-poor medicinal-plant chain (Belt et al. 2003; Alam and Belt 2004).

THE MEDICINAL-PLANT CHAIN IN UTTARANCHAL: FROM COLLECTION TO CULTIVATION

This section describes the current structure of the medicinal-plant chain in Uttaranchal and examines constraints and opportunities for further development of the medicinal-plant chain involving resource-poor farmers in Uttaranchal.

The role of medicinal plants in Uttaranchal

Uttaranchal is one of India's poorest states: in 2001, per capita income was 33% lower than the Indian average (US$ 240). Road and communication infrastructures are not well developed because of the mountainous geography of the area. This limits farmers' access to markets. About 80% of the state's working population depends on agriculture as its main source of (Mountain Technology 2004). As in other parts of the Himalayas, the proportion of land under cultivation is very small. In the plains, about 70% of the total area is cultivated, but only 12% of the total land area of Uttaranchal is under cultivation due to inaccessibility and poor soil quality. Average landholdings are small: more than 50% of the households own less than two acres and only 5% of the households own more than five acres. Furthermore, the average productivity of the region is low and most farmers practice subsistence farming to meet their household needs (Maikhuri et al. 2001). Due to declining returns from traditional crops, farmers in Uttaranchal are only able to survive for 8-9 months a year on farm production. For the rest of the year they depend on non-farm income such as the collection and sale of medicinal plants (Alam 2003). The poor, in particular mainly landless people and marginal farmers, benefit from current collection activities.

Because of its diverse agro-climatic conditions and relative isolation, India's Himalayan region is richly endowed with a large variety of plant species, many of which have medicinal properties. The medicinal plants found in the Himalayan areas include species of particularly high medicinal value (Planning Commission 2000). People in India have long known of the benefits of medicinal and aromatic plants, which provide raw materials for both the pharmaceutical industry and traditional forms of medicine. Besides basic health care, the plants generate income and employment but also have implications for the preservation of biodiversity and of traditional knowledge.

In recent years the demand for medicinal and aromatic plants has grown rapidly because of accelerated local, national and international interest, the latter notably from the pharmaceutical industry in the West. Worldwide, the number of species used for medicinal purposes is estimated at more than 50,000, which is about 13% of all flowering plants (Schippmann et al. 2002). In India, over 8,000 plant species are used in traditional and modern medicine (Planning Commission 2000).

Motivated by the need to increase farmers' incomes through agricultural diversification while conserving biodiversity, the government of Uttaranchal has formulated a special policy to protect medicinal plants and to support commercial cultivation and arrangements for processing and marketing (Government of Uttaranchal 2002). This policy has two main components: regulation of collection of medicinal plants from the wild to protect biodiversity, and promotion of cultivation to meet demand and provide new income opportunities to farmers.

The current medicinal-plant chain: collection from the wild

Most of the medicinal plants from Uttaranchal are collected from forests and rangelands. The State Forest Department is responsible for regulating the collection of species from the wild that are not considered endangered. It determines the areas from which plants can be collected, fixes the volumes to be collected and monitors collection activities in order to prevent illegal and excessive collection. To promote the participation of local communities in conservation activities, the government of Uttaranchal has set up a number of medicinal-plant cooperatives (*Bhaishaj Sangh*). The State Forest Department issues permits to these cooperatives, which in turn employ contractors to organize collection. The contractors employ collectors, usually farmers with small landholdings and landless labourers. The contractors can sell the collected material either to the cooperatives or directly to independent traders after paying royalties to the cooperative. The cooperatives sell to either the local agents of wholesalers, or traders in larger cities or drug manufacturers. The traders supply the domestic market and international markets, mainly in the United States and the European Union (Figure 1).

In the medicinal-plant chain, the collectors and local contractors are in a very vulnerable position. As they cannot sell directly to large traders in big cities, the collectors depend on local traders for market information, credit and the actual marketing of the raw material. This puts them in a weak bargaining position and results in farmers receiving prices that are considerably lower than those prevailing in the wholesale markets. The illegality of the business also puts a downward pressure on prices at the lowest levels in the chain.

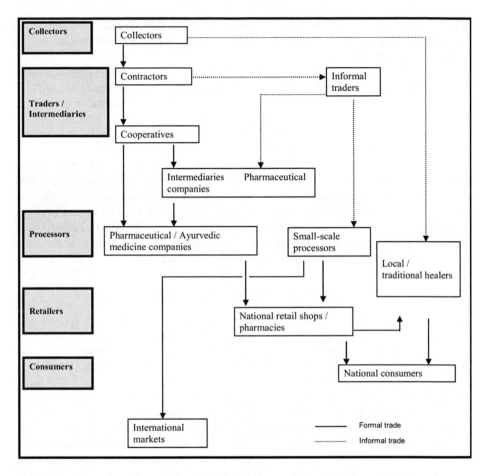

Figure 1. The chain: from collection in the wild to consumption in India (Source: Belt et al. 2003)

The number of local traders, even in the large collection areas, is small. For example, in Munsiyari, a major centre for collection in Uttaranchal, five traders are reported to dominate the trade. Although the number of contractors in Munsiyari has increased to about 20, the trade continues to be dominated by few traders (Virdi 2004). An important reason why contractors and traders exercise such strong control is that the collectors depend on them for loans. As many collectors are poor, they often need to borrow money, which is provided by the contractors and traders. This practice, which is widespread, keeps the collectors tied to local contractors. Also, as they have only small amounts to sell, they do not have the option of selling directly to wholesalers.

In spite of various policy measures, excessive and illegal collection of medicinal plants continues to take place on a large scale. This includes the collection of species considered endangered and whose collection is prohibited by law. The contractors who organize legal collection are often involved in illegal collection as well. As they have connections with both official agencies and large traders, it is easy for them to undertake illegal activities alongside legal trade.

Large-scale collection has led to the depletion of important species in the area. This is reflected in a significant decrease in the amount of material a person can collect in a day. For example, in the Johar valley in the Pithoragarh district, collectors reported that, until five years ago, they were able to collect about 200 grams of dry Atish (*Aconitum heterophyllum*) in one day. Now they do not get more than 70 -100 grams a day (Belt et al. 2003; Alam and Belt 2004).

There are a number of reasons for the excessive collection. Firstly, both collectors and contractors are primarily interested in higher incomes in the short run and have little concern for sustainability. As the contracts are given for only one year, the contractors are primarily interested in maximizing the volume of collection, irrespective of long-term effects. Similarly, the collectors are poor and need to maximize their income to pay back loans taken from contractors/traders. Secondly, the collectors are paid according to volume. Their main interest is to harvest as much as they can in the limited time available to them, irrespective of the consequences. Thirdly, many collectors do not have the traditional knowledge for sustainable collection and have no ownership over the resources they exploit. They use collection methods that are often detrimental to the long-term availability of resources (Belt et al. 2003; Alam and Belt 2004).

Development of the chain: factors limiting medicinal-plant cultivation

Motivated by the need to conserve biodiversity and increase farmers' incomes through agricultural diversification, the government of Uttaranchal has initiated policies to promote the cultivation of medicinal plants. These are being implemented through various government departments such as the Horticulture Department, the Forest Department and the Department of Rural Development, as well as a number of research institutes. Specific measures to promote cultivation include activities to familiarize farmers with the potential of medicinal plants as cash crops; developing and disseminating cultivation technologies; setting up nurseries to propagate and supply planting material to farmers; training farmers; and providing loans and subsidies linked to the cultivation of medicinal plants. Research shows that these policies are yet to have an impact: both the numbers of farmers cultivating medicinal plants and the scale of cultivation remain small in Uttaranchal (Belt et al. 2003; Alam and Belt 2004).

This section describes the main factors that prevent smallholder producers from taking up cultivation of medicinal plants. The factors discussed include the high risks and transaction costs, the lack of trust among chain actors and the need for an enabling institutional infrastructure.

Long gestation period and high risk
Many medicinal plants can be harvested only after three years or more. This is particularly true of the plants grown in high-altitude areas. As most farmers are

poor, have small landholdings and lack credit, they cannot wait so long for returns. Understandably, they are reluctant to convert a significant part of their land to medicinal-plant production.

The cultivation of medicinal plants is also highly risky. This is for a number of reasons. It is a comparatively new activity and reliable cultivation technologies and other inputs are yet to be fully developed. Also, many of the communities currently involved in the cultivation of medicinal plants were traditionally traders. Farming is a comparatively new occupation for them and the risk of failure is particularly high. In addition to the risks of crop failure, the farmers face serious market-related risks and difficulties. Moreover, in most cases they do not have a guaranteed market and price premiums for cultivated material. They also lack reliable market information about demand and pricing, which puts them in a vulnerable position. Local traders often transfer the price risks to them.

Transaction costs
Due to the mountainous geography, the physical infrastructure in Uttaranchal is poor: road networks are not well developed, poor communication networks limit access to information, and agro-processing facilities are limited. For these reasons the transaction costs for rural producers and local entrepreneurs in Uttaranchal are high, even though some of these costs are offset by favourable agro-climatic conditions for the cultivation of high-value medicinal plants and the high value to weight ratio.

Social capital and values
As the medicinal-plant trade based on cultivated material is new in Uttaranchal, various linkages essential for trade are not yet well developed. In the current system the risks of economic coordination opportunism (i.e. risk related to the level of trustworthiness of the actors involved and the chance that arrangements are not respected) are high (Dorward et al. 2004). For example, in the current system traders exert their power to transfer price risks to producers, people often fail to implement agreed actions, and individuals may act opportunistically, withdrawing from collective agreements. Efforts are needed to strengthen the networks of the actors involved in the medicinal-plant chain. Strong social networks (or social capital) can create trust and facilitate cooperation, reducing risks and transaction costs (DFID 1999).

Institutional infrastructure
Being a new state, the institutional infrastructure in Uttaranchal is weak. This is particularly true for institutions that provide technical support and remove marketing bottlenecks. Medicinal plants require specific soil, climate and moisture conditions, as well as interactions with other species, in order to grow. This makes them difficult to cultivate and presents farmers with serious difficulties that they have no experience in solving. There is a clear need to develop technologies related to cultivation, harvesting, storage, transportation and quality control. The state has very

limited infrastructure to generate these technologies. Similarly, the state lacks institutions to provide marketing support to farmers growing medicinal plants. There is also a lack of coordination between various institutions, which diminishes their effectiveness. For example, there is very little collaboration among the research institutes working on medicinal plants, resulting in duplication of efforts and inefficient use of scarce resources. Similarly, there is little collaboration between these institutes, agricultural extension institutions and farmers. This limits both the appropriateness if technology, and its diffusion.

Overcoming constraints by public–private partnerships

Public–private collaborations can play an important role in removing many of the bottlenecks described in Section 2.3. In fact, some promising public–private collaborations have already started in Uttaranchal that aim to overcome some of the existing impediments to the development of a medicinal-plant chain based on cultivated material. This section describes two of these examples, using a contract-farming model: one focusing on the national market and the other on the international market.

PPP: Collaboration between Gheshe farmers, industry and research organization
In Gheshe, a remote village in Uttaranchal, a farmers' organization is involved in a public–private partnership with a national firm and a research institute. The partnership was initiated by the High-Altitude Plant Physiology Research Centre (HAPPRC), which is an important centre of research on medicinal plants in Uttaranchal. Having developed cultivation technology for a number of medicinal-plant species, HAPPRC was searching for modalities to make their technologies available to farmers. It focused on the farmers of Gheshe, with whom it had worked earlier on the cultivation of vegetables. As they had trust in the researchers from HAPPRC, the farmers agreed to start the cultivation of a number of medicinal plants, including *Picrorhiza kurrooa* (Kutki) and *Saussurea lappa* (Kuth). HAPPRC provided seeds and seedlings free of cost as well as technology and training. Following the marketing concerns of farmers, HAPPRC also located a company that could provide a guaranteed market for the production. This resulted in a tripartite arrangement between the farmers' group, HAPPRC and Dhawan International, a Delhi-based firm (Figure 2).

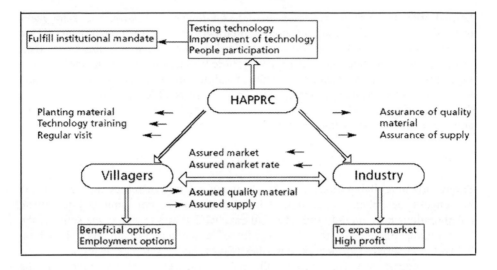

Figure 2. *Collaboration features for public–private partnership in Gheshe village (Nautiyal and Nautiyal 2004)*

The basic conditions of the agreement are as follows:

- The *farmers* cultivate medicinal plants organically and guarantee certain specified quality standards. The farmers are required to sell their produce only to the company. Farmers can ask the company for an advanced loan.
- The *company* guarantees a minimum purchase per growing season at a fixed minimum price. The actual selling price will be based on both the minimum and prevailing price one month before time of delivery. The difference between the minimum price and the selling price is shared equally between the industry and the farmer organization. The price information will be collected by HAPPRC.
- The *research institute* provides technological assistance to farmers to remove any cultivation-related problems and ensure high product quality. The company has the exclusive rights to cultivation based on HAPPRC technology until cultivation covers a minimum area (50ha). HAPPRC is free to transfer the technology to other companies when cultivation extends to more than 50 hectares. HAPPRC charges both the farmer organization and the industry 3 % of the selling price for its services.

As part of this collaborative effort, thirty-two farmers are growing *kutki*. The first harvest of kutki cultivation was taken in October 2004. It produced about 0.2 tons of kutki. A second harvest is planned for May 2005. The tripartite agreement has succeeded in removing some of the bottlenecks in the cultivation of medicinal plants discussed earlier. For example, it provides the farmers with an assured market at a pre-agreed price. This greatly reduces the risk faced by the farmers. The contract also ensures that the farmers will receive planting material, technical support and training from a competent research institute. The industry will receive supplies of cultivated material of a uniform quality, which is not possible in the case of material collected from the forest. The agreement also facilitates the commercialization of cultivation technology developed by public-sector research.

However, it must be pointed out that this collaboration removes only some of the constraints. A number of other issues, such as the need to strengthen the farmers' capacity to collect information on markets and negotiate with industry, are not covered by the agreement. Similarly public–private collaborations do not remove the difficulties created by the lack of appropriate policies and institutions. Civil society and government agencies have important roles to play in building farmers' capacity and improving the efficiency of policies and institutions.

PPPs: matchmaking with international business partners
In an effort to link farmers' organizations from Uttaranchal to buyers in international markets, KIT approached importers of traditional medicines and aromatic plants in Europe to assess the potential for establishing international business linkages. This led to an interest from the Dutch company IHC/VanderStelt. This company imports Ayurveda herbs from India and distributes them in The Netherlands and Germany as health products (capsules and tablets) to pharmacies, chemists, health shops and therapists. Currently, the total product range contains 55 products, all of them based on the Ayurveda principle[1]. Presently IHC/VanderStelt sources its materials from the Covenant Centre for Development (CCD). The latter is an Indian NGO, whose main objective is to promote community enterprise with a focus on the cultivation of medicinal plants. CCD is part of the Foundation for Revitalisation of Local Health Traditions (FRLHT) and produces quality-standard Ayurvedic products. It has approximately 50 hectares of land, where more than 400 species are cultivated. In addition, CCD works with farmers in 300 villages surrounding their unit. It supplies planting material for cultivation and purchases organically produced herbs, from which ingredients are extracted. Since they collect the raw material directly from the growers and make the ingredients, they can guarantee the product quality. IHC/VanderStelt buys directly from the CCD, without the involvement of big industries, ensuring maximum benefit for the 300 communities. It pays a pre-agreed premium on the prices prevailing in India at the time of supply.

The major aim of IHC/VanderStelt is to set up a distribution network in cooperation with the CCD or a similar organization, in order to have a direct distribution channel to therapists all over Europe, with The Netherlands as the gateway to Europe. When production volumes based on sustainable cultivation increase, large parties can be approached, such as pharmaceutical industries and companies that work with natural aromatic substances.

IHC/VanderStelt is interested in procuring organically cultivated medicinal plants from the high altitude areas of India. To make this possible, KIT, CSD and IHC/VanderStelt have agreed to support jointly the organic cultivation of medicinal plants in Uttaranchal. Initially, cultivation will be carried out by about 50 farmers who will receive a guaranteed price for their production, to be exported to IHC/VanderStelt in The Netherlands. The collaboration will enable IHC/VanderStelt to source raw materials from the North of India, where growing conditions for many medicinal plant species are favourable. It is expected that all parties will benefit from the partnership: farmers will have guaranteed access to markets, reduced risk, lower transaction costs, and capacity strengthening of their organizations. KIT and CSD will be able to link IHC/VanderStelt to reliable farmers' organizations and facilitate the cultivation of medicinal plants and their export. IHC/VanderStelt will have access to larger production volumes of organically produced raw materials. It is anticipated that both the number of farmers and the range of medicinal plants to be cultivated will increase as the collaboration progresses.

DISCUSSION AND CONCLUSIONS

At present, resource-poor people in Uttaranchal collect plants from the wild in order to complement their meagre incomes. Due to continued collection and increasing market demand, numerous plant species are threatened with extinction. This has a particularly negative impact on the incomes of the poorest sections of rural societies. For rational and regulated collection, strong local communities or strict governmental control measures are necessary. The first is not in place in Uttaranchal, while collection-control regulations tend to affect the poorest households hardest and push them into illegal, risky ventures. This brings us to the possibility that cultivation of medicinal plants offers a greater opportunity for the poor people of Uttaranchal. It is important that the potential of cultivation of medicinal plants is investigated and the possibilities of public–private collaboration are explored through action research programmes. It is also important to focus this research on issues that affect the livelihood of the poor, including farmers with small landholdings, income opportunities for women, and food security of the poorer section of rural society.

We find that, in spite of considerable government efforts, large-scale cultivation of medicinal plants has not yet taken place in Uttaranchal. It also highlights some of the difficulties that farmers face in carrying out the cultivation of medicinal plants. These include: long gestation period and high risk, poor institutional infrastructure to provide technical and marketing support, high transaction costs and insufficient social capital.

Public–private collaboration is often suggested as a way of reducing these costs to acceptable levels, reduce risk, and secure market access to small producers. Will this make a difference in Uttaranchal and create pro-poor, sustainable chains based on the cultivation of medicinal plants? The example cited in the paper illustrates that such collaborations can overcome many of the constraints and provide a promising mechanism for establishing the conditions necessary for the growth of chains that are in the initial stages of development.

However, the number of public–private collaborations supporting the cultivation of medicinal plants is still very small. What can be done to promote collaboration on a larger scale? A number of conditions will have to be met before the private sector will be attracted to join programmes to support the cultivation of medicinal plants. These include:
1. Public-sector investment, to build the infrastructure necessary for the provision of technical and marketing support

2. Increased involvement of civil society in organising farmers' groups and building capacity to deal with public institutions and private companies, collect market information and build entrepreneurship
3. Build social capital so that the efficiency of the chain is improved. Networks of the actors involved in the medicinal-plant chain must be strengthened both vertically (e.g. producer–industry), as well as horizontally (e.g. strengthen the producer organizations) in order to increase people's trust and ability to cooperate, and expand access to markets. In this process, consulting the stakeholders is not enough. A more profound collective investigation into the motives and underlying values of the stakeholders is essential to enable sustained common action
4. Create greater demand for cultivated material. Presently, the private sector has little reason to participate in joint programmes as it is largely satisfied with the supply of medicinal-plant produce, whether legally or illegally obtained. Only large exporters may be interested in offering cultivation contracts to farmers for species that are difficult to obtain and whose supply fluctuates. Also, cultivated material would be of interest to exporters as it is impossible to trace the origin of collected material, due to a lack of transparency and documentation in the chain. Unfortunately, the role of exports as an impetus to cultivation can only be small for two reasons. Firstly, compared to domestic market, the importance of export is small. This limits their overall influence on the chain. Secondly, it is still possible to export without traceability.
5. The private sector will be more willing to support the cultivation of medicinal plants if the cost of collected material increases significantly. This can happen if the restrictions on collection from the wild are strictly enforced.

Uttaranchal's experience with public–private collaboration to promote the cultivation of medicinal plants by small farmers is at an early stage. It is hoped that these collaborations will provide important lessons that can be replicated in Uttaranchal and other mountainous areas. This would provide strong impetus to agricultural diversification, leading to increased incomes for farmers.

ACKNOWLEDGEMENTS

The authors acknowledge John Belt (SNV Peru), Arjan Lievaart (IHC/VanderStelt), Hugo Verkuijl, Rob van Poelje and Thea Hilhorst (KIT) for comments on the paper. The Dutch Ministry of Foreign Affairs (DGIS) is acknowledged for funding the action-research and development of the paper.

NOTES

[1] Ayurveda is a 4500-year old health-care system, recognized by the World Health Organization

REFERENCES

Alam, G., 2003. *IPRs, access to seed and related issues: a study of the Central and North-Eastern Himalayan region.* Centre for Sustainable Development, Dehradun.

Alam, G. and Belt, J., 2004. *Searching synergy: stakeholder views on developing a sustainable medicinal plant chain in Uttaranchal, India.* KIT Publishers, Amsterdam. KIT Bulletin no. 359. [http://www.kit.nl/net/KIT_Publicaties_output/showfile.aspx?a=tblFiles&b=FileID&c=FileName&d=TheFile&e=605]

Belt, J., Lengkeek, A. and Van der Zant, J., 2003. *Cultivating a healthy enterprise: developing a sustainable medicinal plant chain in Uttaranchal-India.* KIT Publishers, Amsterdam. KIT Bulletin no. 350. [http://www.kit.nl/publishers/assets/images/isbn9068328395_compleet.pdf]

DFID, 1999. *Sustainable livelihoods guidance sheets.* Available: [http://www.livelihoods.org/info/info_guidancesheets.html] (2004).

Dorward, A., Kydd, J., Morrison, J., et al., 2004. *Institutions, markets and policies for pro-poor agricultural growth.* Centre for Development and Poverty Reduction, Imperial College Wye, Wye. [http://www.imperial.ac.uk/agriculturalsciences/research/sections/aebm/projects/poor_ag_downloads/ghentpap2.pdf]

Dr.J.VanderStelt, 1999. *Ayurveda.* Available: [http://www.drvanderstelt.nl/] (2004).

Government of Uttaranchal, 2002. *Marketing of medicinal plants: status and action plan.* Government of Uttaranchal, Horticulture and Rural Development Department, Dehradun.

Maikhuri, R.K., Rao, K.S. and Semwal, R.L., 2001. Changing scenario of Himalayan agro-ecosystems: loss of agrobiodiversity, an indicator of environmental change in Central Himalaya, India. *The Environmentalist,* 21 (1), 23-39.

Mountain Technology, 2004. *Mountain farming system.* Available: [http://mountaintechnology.tripod.com/intro/mtnfarmsys.html] (2004).

Nautiyal, M.C. and Nautiyal, B.P., 2004. Collaboration between farmers, research institutions and industry: experiences of Picrorhiza kurrooa cultivation at Gheshe village in Chamoli district, Uttaranchal. *In:* Alam, G. and Belt, J. eds. *Searching synergy: stakeholder views on developing a sustainable medicinal plant chain in Uttaranchal, India.* KIT Publishers, Amsterdam, 63-72. KIT Bulletin no. 359.

Planning Commission, 2000. *Report of the Task Force on Conservation and Sustainable Use of Medicinal Plants.* Planning Commission, New Delhi. [http://planningcommission.nic.in/aboutus/taskforce/tsk_medi.pdf]

Schippmann, U., Leaman, D.J. and Cunningham, A.B., 2002. Impact of cultivation and gathering of medicinal plants on biodiversity: global trends and issues. *In:* Inter-Departmental Working Group on Biological Diversity for Food and Agriculture ed. *Biodiversity and the ecosystem approach in agriculture, forestry and fisheries: satellite event on the occasion of the ninth regular session of the Commission on Genetic Resources for Food and Agriculture, Rome, 12-13 October 2002.* FAO, Rome, 1-21. [ftp://ftp.fao.org/docrep/fao/005/aa010e/AA010E00.pdf]

Virdi, M., 2004. Wild plants as resource: new opportunities or last resort? Some dimensions of the collection, cultivation and trade of medicinal plants in the Gori basin. *In:* Alam, G. and Belt, J. eds. *Searching synergy: stakeholder views on developing a sustainable medicinal plant chain in Uttaranchal, India.* KIT Publishers, Amsterdam, 41-54. KIT Bulletin no. 359.

SUMMARY AND CONCLUSIONS

CHAPTER 17

SUSTAINABLE AGRO-FOOD CHAINS

Challenges for research and development

LOUISE O. FRESCO

FAO, Vla della Terme di Caracalla, 00100 Rome, Italy. E-mail:
louise.fresco@fao.org

Abstract. The current paradigm shift from supply-oriented and quantity-driven agriculture towards an understanding of demand- and market-driven food chains is leading to greater attention for quality aspects and environmental concerns. Rapid urbanization and increases in purchasing power cause important changes in dietary patterns that will restructure agriculture in the world. This affects the food chain in terms of the types of products that are produced and the kind of research that is required. For meeting global food demand, the volume of agricultural production will have to double through a process of sustainable intensification. Food safety has become a key factor in trade negotiations and bio-security is absolutely essential to all consumers in the world. Implementing quality standards begins with the selection of farmers' production methods. Knowledge institutions should face the challenge to design products and processes that meet food demands within developing countries while considering their specific ecological conditions.

Keywords: agricultural development; dietary patterns; bio-security; trade barriers; food safety standards

INTRODUCTION

It is obvious that we have not yet reached a full understanding of all aspects of international agro-food chains. In this brief final chapter, I cannot do full justice to the richness of the discussions of which this volume is a reflection. Therefore, I will only present here some of my thoughts and ideas, keeping in mind that my daily work is in FAO, one of the biggest international organizations dealing with agro-food issues.

SHIFTING PARADIGMS

We seem close to a paradigm shift in the way we look at agricultural and rural development. The conference that laid the foundation for this book would not have been possible five years ago. We are moving away from a view of agriculture and food that used to be highly supply-driven, quantity-driven and in fact mainly cereals-driven, towards a paradigm encompassing the entire food chain and

R. Ruben, M. Slingerland and H. Nijhoff (eds.), Agro-food Chains and Networks for
Development, 205-208.

including environmental concerns. This new paradigm is strongly demand-driven, market-driven in terms of involving many segments of the market. Hence, it is quality-driven and not just quantity-driven, and it relates to a very diversified and volatile type of market. In other words, our view of agriculture and food chains is diversifying and becoming more complex.

In twenty years from now, more than 60 percent or more of the world's population lives in cities. This rapid urbanization and the related income increase cause important changes in dietary patterns that will completely restructure agriculture in the world. We can expect a shift towards more fruit and vegetable consumption amongst the middle and upper middle classes, and more animal-protein consumption across all levels. The FAO/WHO report on diet, nutrition and chronic health indicates that changes in dietary patterns will be tremendous. The greatest increases in purchasing power will be felt in the group of 20 rapidly developing countries (Brazil, India, China, South Africa and others). Beyond large changes in dietary patterns we may expect attempt to match individual health and diet, leading to an increasing awareness of functional foods. That will affect the food chain in terms of the types of products that will be produced and the kind of research that is necessary.

AGRICULTURAL DRIVERS OF GROWTH

Seventy percent of the poor live in rural areas. For them agriculture will continue to be the main driving force for development. In many developing countries, the main basis of the economy is and remains to be agriculture. There are only two economic sectors that are likely to grow in these countries, particularly in sub-Saharan Africa, namely the energy sector and the agricultural sector. They provide most employment and will grow at the rhythm of population growth, even if nothing else changes and even if there is no export. We may see changes in these two sectors and they may become even more interconnected. Not only because agriculture can supply some bio-fuels, but also because plants are far more efficient producers of basic primary inputs for the chemical industry than the oil industry. We need to be aware of these changes now, because of the time lags of research and research investments. Overall, the volume of agricultural production will have to double to meet rising demands, irrespective of a quality shift in dietary patterns. That growth is in itself already an important challenge. All additional production has to come from the same natural-resource base we currently have available. The only real options for satisfying growing demand is through a process of sustainable intensification.

GLOBALIZATION, LIBERALIZATION AND SMALLHOLDERS

The core question regarding agro-food chain development is: how can small producers, small countries and – I would add – small companies benefit from this? The world does not only exist of a few large retailers. There are many small companies, both at the demand side and at the supply side. Innovation – particularly in Asia – takes place in small family-based companies that distribute seeds, produce tools, process foods etc. The available room for smallholders depends on how they can benefit from liberalization and globalization. Although these terms are sometimes used as more or less interchangeably, they are not at all the same. Globalization is a process that can be loosely defined as an increased movement of people and goods around the world. One of the central problems in globalization is that we are moving germs, not just human germs but also many pathogens. That is the reason why food safety has become such an essential key factor nowadays in trade negotiations. I really would like to stress here that food safety – safety in the food chain – or what we more broadly at FAO define as bio-security, is not a luxury for rich consumers or rich countries. Bio-security is absolutely essential to all consumers in the world.

Liberalization has to do with prices and price support that affects the world market price. As far as I can see, in the next few years these price barriers will certainly go down and are bound to disappear. The latest developments in the Doha round suggest that this is only a matter of time. The potential effects are complex and may not immediately benefit the poor. For example, the liberalization of cotton prices is likely to benefit China, India and to a minor extent Egypt, but not one single African country. Therefore, liberalization of prices is no panacea. More importantly, the greatest barriers of trade are not between the developing countries and Europe or the US, but exist between developing countries themselves. There are major customs and trade barriers that actually prohibit interregional trade. In any case, a large share of the growth will come from south-south trade, south-south cooperation and south-south companies working together.

STANDARDS

The use and application of standards in trade is currently a matter of much debate and confusion. There are different types of standards and various types of legal arrangements. First of all, we have the so-called *Codex Alimentarius* standards on food safety. In view of the globalization mentioned earlier the advantages of having universally uniform food standards for the protection of consumers are self-evident. The Agreement on the Application of Sanitary and Phytosanitary Measures (SPS) and the Agreement on Technical Barriers to Trade (TBT) both encourage the international harmonization of food standards. A product of the Uruguay Round of multinational trade negotiations, the SPS Agreement cites Codex standards, guidelines and recommendations as the preferred international measures for facilitating international trade in food. As such, Codex standards have become the benchmarks against which national food measures and regulations are evaluated as well as others that are formally recognized under the WTO.

However, concerns on food safety evolve faster than the standards adopted by Codex. Some 'standards' are set in a voluntary manner, either by the private sector, by countries, or by groups of consumers on ethical grounds. They may have similar effects on the market access of small farmers and business in developing countries. Although it has been suggested to develop special standards for developing countries, the way to go is to invest in capacity building to enable developing countries and small farmers to participate in negotiations.

Food chains are very, very complex, involving many actors and many steps. When talking about partnerships, alliances have to be established in every single step. They start from base level, from individual farmers, all they way up to the international market and consumers. There is no such thing as linking the small farmer directly to the international market. Although the conference has said little about NGOs, they may also contribute in forging in these partnerships. Public resources and development-bank funds may be used as a guarantee to start this first stage of collaboration.

CHALLENGES AHEAD

What role can knowledge institutions like universities play in supporting sustainable agro-food chains and networks? First, I expect that the agro-food chain approach needs to be reflected in the curricula we teach and in the research we undertake. In fact, I believe that nobody should be able to graduate without a basic understanding of how agricultural and food markets work. Even students graduating in other areas should recognize the essential elements of how agricultural trade in the world is organized. In addition, we need more research on new products to respond to future demands for differentiation. The challenge is to design products and processes that meet these demands within the specific locations of developing countries and considering their specific ecological conditions.

Finally, I would like to stress the current and future importance of environmental issues. These are directly related to quality performance. Some remarks were made that quality is not an issue that small farmers can do something about, but that is not true. Take standards on maximum residue levels of pesticides: farmers' production methods make all the difference. Being aware of the environmental aspects in every single step of the chain, should be part of the way we think about rural development and development cooperation. Agro-food chains will become an ever more important part of our thinking: of negotiations, of our teaching and of our research.

CHAPTER 18

EXCLUSION OF SMALL-SCALE FARMERS FROM COORDINATED SUPPLY CHAINS

Market failure, policy failure or just economies of scale?

CORNELIS (KEES) L.J. VAN DER MEER

Agriculture and Rural Development Department, World Bank, Washington DC, USA. E-mail: cvandermeer@worldbank.org[1]

Abstract. Coordinated supply chains are rapidly increasing in importance in global food markets. They are commercial tools for competitive strategies, assuring quality, food safety and better logistics. They serve high-end markets, especially in industrial countries, but increasingly also in developing countries in urban areas with relatively high incomes. However, the share of production in developing countries marketed through coordinated supply chains is still small. There is widespread fear that small-scale farmers will be excluded from coordinated supply chains. Empirical evidence is mixed; there are abundant examples of successful inclusion as well as of painful exclusion. In some cases, economies of scale are such that only large-scale enterprises can compete successfully in global markets. But, in many other cases there is no level playing field. Analysis of factors that contribute to inclusion and exclusion indicates that there are market failures and policy failures contributing to relatively weak competitiveness of small-scale farmers. Hence, public intervention can be warranted.
Keywords: food safety; incentives; public sector; private investments; farmers' organization; risks; contracts

TRENDS IN FOOD MARKETS

Rapid changes have taken place in global food markets in recent years. Changes in consumer demand, food safety concerns and the rise of modern retail systems are the main drivers for these changes. With higher income and changing lifestyles, demand has increased for more variety, higher quality, year-round supply of fresh produce, 'healthy' food, convenience and value added. There is a rapid increase of demand for 'ready-to-eat' food. And, last but not least, consumers require safe food, and they have increasing concerns about the social and environmental conditions under which food is produced.

Food industries, supermarkets and food services compete for the market shares and market power by trying to meet consumers' preferences. They have become important buyers in global markets and ask for specifications that meet consumer

R. Ruben, M. Slingerland and H. Nijhoff (eds.), Agro-food Chains and Networks for Development, 209-217.

demand. These private-sector specifications are often much more demanding than public-sector requirements of food safety and quality. Food industries serve consumers with attractive processed products. Supermarkets try to offer an attractive assortment of products at a one-place-to-shop. Food services – restaurants, canteens, fast-food outlets – offer direct service to consumers, which by-passes supermarkets. The market share of food services is growing faster than that of supermarkets in many countries.

Information technology, logistics and advances in food processing and post-harvest handling have greatly enhanced the development of global sourcing and retailing. Trade liberalization has contributed to a rapid growth of international trade in food, especially for fruit, vegetables and fisheries products (Diop and Jaffee 2005; Hallam et al. 2004).

Economies of scale are important in retailing, transport, logistics and processing, and there is a clear concentration among retailers, food services and food processors. Yet, there is heavy competition from which companies try to escape through strategies of product differentiation, branding of products, and product and market innovation.

Food safety concerns have been an important accelerator of changes in food market development. Many countries have seen food scandals and food scares. Most important examples are BSE, high residues of pesticides and antibiotics, dioxin and toxic chemicals in the food chain, *Listeria*, *Salmonella* and other microbiological hazards, hepatitis, and recently, avian influenza. These scandals and scares received major attention in the media and contributed to consumers' concerns. It is fair to say that consumers have become more suspicious about the trustworthiness of food regulators, scientists and the food industry, and in many countries this translated into political pressure to strengthen public control. Fears of bio-terrorism have added to this. As a result, food laws and regulations have been revised in Japan, the EU, the USA and elsewhere, responsibilities have been sharpened, and border controls intensified.

The private sector has been much affected by food safety crises over a decade or so. It sometimes experienced heavy losses because of stocks that had to be discarded, interrupted supply, loss of business, and damage to company and brand name. A number of cases resulted in bankruptcies. Nowadays, most food companies treat food safety as an important commercial risk, but also as a subject with opportunities to distinguish themselves from competitors. They deal with food safety risks through increased control of the supply chains from farm to table. They abandon open markets with anonymous suppliers and instead turn to integrated or coordinated supply chains. This usually involves reliance on preferred suppliers who assure safety through tracking and tracing, and independent certification of good agricultural and good manufacturing practices.

These trends in consumer demand, retailing and food safety management are most visible in the high-income industrial countries, but in developing countries the same trends can be observed in urban areas with relatively high incomes, although the impact levels are still lower.

COORDINATED SUPPLY CHAINS

Coordinated supply chains are durable arrangements between producers, traders, processors and buyers about what and how much to produce, time of delivery, quality and safety conditions, and price. They often involve exchange of information, and sometimes also help with technology and finance. They are usually initiated by investment of private traders and food companies, who act as chain leaders. They have characteristics of partnerships and joint interest. By contrast, relations in open supply chains are usually limited to transactions only. There are hardly contractual relations and little clear loyalty between buyers and sellers. In fully integrated supply chains, on the other end of the spectrum, one company performs all activities from production to processing and wholesaling on its own account without partnering with other entities. Here the intra-firm handling has replaced market transactions. Coordinated supply chains as an institution have to compete with atomistic markets on the one hand and with the firm that completely controls the supply chain on the other.

Coordinated supply-chains well fit the logistics requirements of modern food markets, especially those for fresh and processed perishable food. They can be used for process control of safety and quality. This is more effective and efficient than only control at the end of the supply chain. Companies cannot control each single package that is sold; they need total quality and safety management as well. Companies use coordinated supply chains also as tools in competitive strategies, such as for sales promotion, labelling and branding.

With the emergence of coordinated supply chains, competition is increasingly between supply chains rather than between individual firms.

In the past ten years, coordinated supply chains spread rapidly in food markets of the industrial countries. The spread mainly depended on increase of sophistication in consumers' demand, stringency of quality and safety requirements, and the possibilities of efficiency gains through improved logistics. This means that penetration is highest in the market segments that cater for the top-end retailers in industrial countries. It is higher for perishable products than for staples, because of the higher food safety risks. Certain product and market segments in the food-processing industries have higher risk and vulnerability than others, such as baby food and dairy products. They use stringent supply-chain control for managing risk. Restaurants and other fast-food chains are also very vulnerable to food scandals, hence, coordinated supply chains play prominent roles. In bulk product segments with lower quality and safety risk, coordinated supply chains play modest roles at best.

In developing countries the pattern is similar to that in industrial countries, but the level of penetration of coordinated supply chains is much lower. Coordinated supply chains are relatively widespread in the small production segments that cater for demanding export markets, especially those for perishable products and sensitive processed products destined for industrial countries. They are emerging slowly in the perishable products for the domestic supermarket segments, international hotels, modern restaurants, and food processing. They are virtually absent in the large traditional food and commodity markets.

Although coordinated supply chains are spreading rapidly, in both more and less developed countries, the share of small-scale farmers in developing countries affected by them is still small. An example is the spread of coordinated supply chains in China. The export share of vegetables is about 1 percent of the volume of production, and of fruits about 2 percent (*The China statistical yearbook*; *China customs data*). The share of fresh fruit and vegetables that goes through supermarkets in Shanghai, the country's most developed city, is less than ten percent and a significant part of that product is still sourced from open wholesale markets (the author's interviews). This means that for the country as a whole the share of fruits and vegetables that goes through coordinated supply chains is a few percent at most.

Contract farming arrangements are coordinated supply chains or parts thereof, and while much of the literature on contract farming is relevant to coordinated supply chains, for brevity it is not addressed explicitly here (Baumann 2000; Eaton and Shepherd 2001).

What are the incentives for investing in coordinated supply chains?

Why would enterprises – and farmers – invest in the formation of coordinated supply chains? Why would a trader or processor try to become a chain leader? Advantages can be that better prices can be obtained. Supply-chain control may help in achieving higher quality and safety standards, which command better prices. It may result in getting or maintaining access to higher-end markets, which pay better. It may be a pre-condition for successfully adding value to primary products, such as cleaning, packing for supermarkets or processing. Advantages can also be reduction of cost through higher efficiency and reduction of losses. In particular, in fresh produce losses can be high because of uncoordinated supply and limited shelf life. Transaction costs may be lower if the supply chain is shortened by bypassing traditional markets. Demand for regular daily supply of fresh product or supply for retailers or processors can only be met through coordination in the supply chain, and the supplier who can organize that successfully will have a stronger market position. This is closely related to achieving more economies of scale and scope. And, last but not least, managing risk successfully can also require coordination between producers, traders, exporters and buyers. Often advantages include a mix of these benefits. But, benefits come with costs and the decision to invest in a coordinated supply chain is a commercial one. It is based on assessments of costs, benefits and risks.

Can small-scale farmers participate in coordinated supply chains?

There is a widespread concern among development specialists that small-scale farmers are excluded from coordinated supply chains. This fear has been strengthened by several studies led by Tom Reardon on the rapid growth of supermarkets in developing countries, which point at drastic impacts for and exclusion of small farmers from supermarket supply chains; Reardon and Berdegué (2002) is one of the leading studies in this field.

Small-scale farmers have particular strengths and weaknesses for participating in high-end markets, compared to commercial farms. The important question is whether supply chains with small-scale farmers can compete successfully with supply chains with commercial farms and with integrated supply chains.

The main strength of small-scale farmers, which comes up frequently in interviews and literature (interviews by the author and, for example, Baumann 2000, p. 31) is that their production cost in labour-intensive products is often 20-40% lower than that of large-scale commercial farms. The latter have high overhead and supervision costs and paid labour is generally less motivated than are self-employed farmers. In some cases, where lack of access to land forms an obstacle to the emergence of commercial farms, access to land is a competitive strength of small-scale farmers.

Important weaknesses of small-scale farmers are lack of knowledge about modern markets, modern technology and proper use of modern inputs. Access to capital can be an obstacle for upgrading from production for direct consumption and local markets, rather than to more demanding markets. Working with small-scale farmers is difficult for trading and processing companies. Quantities of product are small and heterogeneous in quality, supply can be haphazard and bulking-up of volume into a steady stream of product of constant quality difficult to realize. These are serious problems for serving high-end modern supply chains. The organization of small-scale farmers is not easy in many cases. The culture of existing cooperatives and organizations may be an obstacle rather than an asset. Enterprises interested to step in and help with the organization of production and marketing face high transaction costs, and contract enforcement that is costly or even impossible. Risks of working with small-scale farmers can be high because of their ignorance and higher incidence of inappropriate application and use of illegal agrochemicals.

Without support from traders and processors small-scale farmers will rarely be able to overcome their weaknesses and participate in supply chains for high-end markets. Traders and processors will include small-scale farmers in coordinated supply chains only if they expect the benefits to outweigh costs (Box 1 has an example)[2]. Important factors are the perceived benefits, costs and risks.

The risks will depend on many factors. With extensive markets for cheap illegal pesticides and illegal antibiotics, the risks for uses of illegal pesticides and antibiotics by small-scale farmers are higher than in cases of effective public control. Government support for educating farmers in proper use of agrochemicals will also help in reducing risks. Risk of working with small-scale farmers on complex technologies will be relatively high. Given the potential risk with contract enforcement and loyalty, government culture matters. A culture of disrupting government interference in markets, debt forgiveness and weak contract enforcement will result in higher risks of working with small-scale farmers. Increased mutual trust between farmers and enterprises will reduce the perceived costs and, last but not least, good organization of farmers and effective leadership is a crucial factor in overcoming many of the weaknesses.

Benefits of working with small-scale producers in coordinated supply chains will be highest for labour-intensive products. An enterprise may achieve benefits of a coordinated supply chain through increased economies of scale in the market, since more volume of consistent quality may allow for better contracts with buyers. Experience in working with small-scale farmers is an important factor for success in enterprises that seek to link to such farmers. There is a need to develop suitable contractual relations and benefit sharing, with the aim to achieve loyalty and reduced costs in the supply chain. Farmers participating in coordinated supply chains receive better prices than they could obtain in open markets. Loyalty is a crucial variable for sustainability and will depend on higher prices that can be paid or on high shifting cost for farmers who want to divert to other buyers. Experience among farmers in working with enterprises can result in reduced transaction cost. This will always require well-organized groups with good leadership. Various activities at the local level can be much better performed through self-control by farmer groups, and through farmer leaders rather than by enterprise staff.

Box 1. A rewarding pro-active strategy

A Thai packing house that collected horticultural products from small-scale producers and delivered packaged products for export to an exporter, received strong signals in the late 1990s from buyers in the UK that it had to upgrade to the new retail standards, BRC and EUREPGAP. The company decided to pursue a pro-active and offensive strategy. The company acquired land on which to establish good agricultural practices. It upgraded all its facilities, introduced HACCP and ISO 9001 with external certification and was BRC-accredited in 2003. Its farm is EUREPGAP-accredited. It has heavily invested in the training of its staff. The company shortened the supply chain by exporting directly and leaving out the exporter. The company has long-term daily delivery schedules with buyers, which enabled the company to negotiate low airfreight rates.

In 2003, the company produced about 35 per cent of the value of its shipping on its own farms, but that share is declining. The rest it buys from small-scale farmers through a system of contract arrangements through brokers with farmers' groups and their individual members. The brokers provide technology and ensure compliance with delivery requirements. For vegetables, Good Agricultural Practices (GAPs) are prescribed, and growers receive training, seeds, pesticides and other inputs. The inputs are repaid in kind. Use of inputs and production are registered daily. The farmer groups have first responsibility to control compliance with GAP requirements themselves. Farmers receive prices 20 percent higher than in local markets. The number of contract farmers was about 900 at the end of 2003 and increasing. The company's sales increased from US$3.3 million in 1999 to US$8.5 million in 2003 and an estimated US$11.5 million in 2004.

(Source: Interviews by Sompop Manarungsan and Kees van der Meer in May and November 2004)

Cost for investors in setting up and operating coordinated supply chains will be reduced if the investment climate is good – infrastructure, contract enforcement, public and commercial services – and if there is public support for high set-up costs for training and for the development of applied technology.

POSSIBLE POLICY INTERVENTIONS

Coordinated supply chains are commercial tools in competitive strategies. They are institutional arrangements that help the private sector, including farmers, to achieve net benefits that otherwise would not be available. Their creation depends on investment decisions, mainly by the chain leader, but also by farmers and others involved in the partnership. Once parties have invested, they have a joined interest. Breaking up a coordinated supply chain results in loss of capital. Hence, there is often loyalty and durability in partnerships that form coordinated supply chains. Needless to say, arrangements in a coordinated supply chain require a different culture from that prevalent in the transaction supply chains. Competition in the high-end markets is increasingly competition between (integrated and coordinated) supply chains rather than between individual enterprises. It is also competition between coordinated supply chains and small-scale farmers and commercial farmers.

Although coordinated supply chains are commercial arrangements in which the private sector must be in the driver's seat, there are a number of things the public sector can do to enhance their emergence and functioning. These things are related to the investment climate and involve public-goods aspects. Four fields for public intervention deserve to be mentioned. First, the government must provide adequate laws, regulation and enforcement necessary for doing business, in particular in food supply chains in which small-scale producers are involved. Important areas of attention are regulation of markets for pesticides and veterinary drugs. But also property rights and contract enforcement deserve attention. Second, there is a role for independent facilitators – honest brokers – to help in overcoming lack of experience and lack of trust among enterprises and farmers. This can be done by contributing well-documented information about arrangements and experiences that have worked elsewhere, and tailoring arrangements to specific needs.

A third area is the promotion of producers' organizations. In many countries legislation and regulation discourage the formation and development of independent producer organizations. Memories of failed policies that initiated or supported public-sector-dominated cooperatives contribute to negative attitudes among farmers and others about farmers' organizations. Subsidization and debt forgiveness schemes of the past have contributed to a wrong business culture among farmers. Governments must clearly break with policies of the past in order to give farmers a good chance as partners in coordinated supply chains. Support for independent producers' organizations, training of leadership and education about modern markets can be helpful. Fourth, support can be given for the development of good agricultural practice (GAP), good manufacturing practice, improved technology and training. Often it may be a good strategy to let the chain leader take the lead, since the results are directly related to market success and provide some sharing in the cost.

FINAL REMARKS

The market served by coordinated and integrated supply chains is more visible than traditional markets, but in most countries it constitutes still a relatively small share of production, often a few percent only. It is the more profitable part of the agricultural sector and it caters for the export market of perishables and vulnerable processed products. It is also an emerging but small part of the modern domestic retail market. It gets a relatively large amount of attention given its modest share in production and the limited involvement of poor farmers and labourers.

Coordinated supply chains are spreading rapidly in high-end food markets. There is much concern whether small-scale farmers can participate in these relatively profitable parts of the food markets. There is evidence of successes from many parts of the world. Many of these successes occurred without or with little public support. But, there are also examples of exclusion; cases where small-scale farmers lost out in competition from large-scale competitors. Rapidly increasing requirements for food safety and quantities of consistent quality contribute much to rapid restructuring of supply chains. It is important to analyse and understand reasons for success and failure, and to design public intervention in cases where it is warranted.

Global food markets are characterized by economies of scale and scope, especially in logistics, marketing and technology. Small-scale farmers, even if they are well organized and not discriminated against, cannot be competitive in all products and markets. There is nothing wrong with that. In such cases large-scale enterprises can be competitive and in so operating generate on-farm and off-farm employment in serving high-end markets. However, in cases where the growth of large-scale companies is based on neglect or discrimination of small-scale farmers rather than on economies of scale there is a need to create a level playing ground. Small-scale producers may be in a weak position because of market failure or policy failure.

There are *market failures* that bear relatively heavily upon small-scale producers and that can put them in a disadvantageous position for participating in coordinated supply chains. Small-scale farmers are often poorly organized, and risks and transaction costs of involving them in coordinated supply chains are relatively high. Small-scale farmers are more affected by negative externalities of markets for agrochemicals and by problems of moral hazard than are larger scale farmers. Markets that provide small-scale farmers with information and technology are often incomplete and inefficient.

There may also be *policy failure*, which can put small-scale farmers in a weak position. An important policy failure is failing to mitigate problems of well-understood market failure. Failure to properly control markets for agrochemicals, for instance, negatively affects the competitiveness of small-scale farmers. Producer organizations have to play important roles in reducing transaction costs and risks of working with small-scale producers. However, inappropriate policies inhibit or discourage their development. Support provided for development of independent producer organizations and applied technology is often insufficient. A culture of political interference in markets and poor contract enforcement increases the risk for enterprises of working with small-scale farmers.

This note provides a framework for understanding the rapid spread of coordinated supply chains and why small-scale farmers are included or excluded. It also provides arguments for policy intervention in cases where there is no level playing field for small-scale and large-scale producers because of market and policy failure. The World Bank's Agriculture Investment Sourcebook provides some guidance for work in this field (World Bank 2004). There is much need for more empirical research for better understanding of economies of scale and of market and policy failures in this domain, which includes a wide range of products, local situations and markets. Such analysis will help to better inform policy development and intervention in the diverse circumstances being faced by the World Bank's clients.

NOTES

[1]. This paper was orally presented at the workshop "Is there a place for Smallholder Producers in Coordinated Supply Chains?", World Bank, Washington, DC, December 8, 2004. The author benefited from helpful comments from Jock R. Anderson
[2]. Swinnen, in a World Bank study (Swinnen 2004) and other publications, documents many examples of successful coordinated supply chains between small-scale farmers and agribusinesses in former communist countries of Central and Eastern Europe

REFERENCES

Baumann, P., 2000. *Equity and efficiency in contract farming schemes: the experience of agricultural tree crops*. Overseas Development Institute, London. Working Paper no. 139. [http://www.odi.org.uk/publications/wp139.pdf]

China customs data. China Customs Statistics Information Service.

The China statistical yearbook. China Statistics Press. Various issues.

Diop, N. and Jaffee, S.F., 2005. Fruits and vegetables: global trade and competition in fresh and processed product markets. *In:* Aksoy, M.A. and Beghin, J.C. eds. *Global agricultural trade and developing countries*. World Bank, Washington, 237-257. Trade and Development Series. [http://siteresources.worldbank.org/INTGAT/Resources/GATfulltext.pdf]

Eaton, C. and Shepherd, A.W., 2001. *Contract farming: partnerships for growth: a guide*. FAO, Rome. FAO Agricultural Services Bulletin no. 145. [http://www.fao.org/ag/ags/AGSM/contract/cfmain.pdf]

Hallam, D., Liu, P., Lavers, G., et al., 2004. *The market for non-traditional agricultural exports*. FAO, Rome. FAO Commodities and Trade Technical Paper no. 3. [http://www.fao.org/docrep/007/y5445e/y5445e00.HTM]

Reardon, T. and Berdegué, J.A., 2002. The rapid rise of supermarkets in Latin America: challenges and opportunities for development. *Development Policy Review*, 20 (4), 371-388. [http://www.blackwell-synergy.com/doi/pdf/10.1111/1467-7679.00178]

Swinnen, J.F.M., 2004. *The dynamics of vertical coordination in ECA agrifood chains: implications for policy and bank operations: final report*. World Bank, Washington. Report no. 29457. [http://www-wds.worldbank.org/servlet/WDS_IBank_Servlet?pcont=details&eid=000011823_20050411110737]

World Bank, 2004. *Agriculture investment sourcebook*. World Bank, Washington. [http://www-esd.worldbank.org/ais/]

CHAPTER 19

FOOD CHAINS AND NETWORKS FOR DEVELOPMENT

Lessons and outlook

MAJA SLINGERLAND, RUERD RUBEN, HANS NIJHOFF AND PETER J.P ZUURBIER

Wageningen University and Research Centre, P.O. Box 9101, 6700 HB Wageningen, The Netherlands

Abstract. Agro-food chains and networks can become an important instrument for development, particularly if smallholder participation can be guaranteed and adequate support is provided for capacity development and upgrading. These conclusions can be derived from the presentations delivered at the international conference organized by Wageningen University , The Netherlands, on 6 and 7 September 2004. Keynotes from representatives of public agencies, nongovernmental and farmers' organizations, scientists and different chain partners (from farmers to retailers) – both from the South and the North – identified a number of strategic policy issues that deserve attention. Different business cases offered a rich range of experiences, empirical evidence and lessons learned for successful supply-chain integration. The conference's main aim was to build bridges between scientific research and development practice. In this final contribution, major challenges for research and feasible options for interventions are identified that can contribute to developing integrated agro-food chains and networks and to improving their added value as a pathway towards pro-poor and sustainable development.
Keywords: critical success factors; supply chain development strategic research; policy agenda.

INTRODUCTION

The food-chain and network approach focuses on jointly enhancing the performance of farmers and companies involved in the agricultural sector and in the agribusiness food and retail industries. Traditionally, smallholders amidst input suppliers and buyers are often perceived as the weakest link in the food chain, due to their small scale and limited negotiating power. Where there is an abundance of agricultural commodities in the global market, causing downward pressure on the prices paid to farmers, power has shifted downstream in the food chain. Moreover, where public policy is geared to lifting protectionism nationally or regionally, farmers are facing the cold wind of competition, urging them to set out survival strategies. However, it is a misunderstanding that due to the shifting of power, food-manufacturing

R. Ruben, M. Slingerland and H. Nijhoff (eds.), Agro-food Chains and Networks for Development, 219-231.

companies, traders and retailers are automatically benefiting and capturing all the rents. In many cases (see the presentations by Schmid of Royal Ahold and Van Deventer of Shoprite) these actors in the food chain also face severe competition. Chain reversal, shaping supply chains in response to consumer demands, leads to competition on price and product quality. Companies therefore have to earn their 'licence to sell'. In addition, civil-society organizations formulate additional demands aiming at environmental and social objectives linked to the organization of production processes and thus underpinning the 'licence to produce'. Focussing on agro-food supply-chain integration and upgrading can therefore be envisaged as a promising approach for reconciling both mandates and to assist chain partners in developing a sustainable competitive advantage (as discussed in the introduction by Ruben et al.).

Current tendencies of increasing concentration in the global food retail sector and in the food-manufacturing industry (Reardon and Timmer in press) ask for complementary strategies to assist all stakeholders involved in the supply chain in improving competitiveness and building-up balanced negotiating power. From these points of view, it is not remarkable that farmers, cooperatives and food companies are all looking for new opportunities to expand business domestically, within the region or in North-South directions in a way that fits consumer and citizen interests. The Wageningen conference identified several of these strategies and addressed key questions like:

• Can cross-border agro-food chains and networks make a difference?
• If yes, what are critical factors to enhance their success?
• How to organize international agro-food chains and networks?
• How to provide mechanisms for sustainable food chain integration?
• What are the challenges for research and development to support cross-border food chains and networks?

In the current development discourse, agro-food chains are sometimes forwarded as a pathway for alleviating poverty, to promote equity (i.e. gender, black power) and to contribute to environmental sustainability in the South. Citizens in the western world may acknowledge these dimensions of agro-food chain development if they express their concerns through nongovernmental organizations (NGOs). Industries and retailers also attempt to address some of these aspects within the framework of their Corporate Social Responsibility (CSR). Supply-chain integration for development, however, aims at mainstream criteria of equitable sharing of benefits and reducing externalities through the participatory development of production, processing, handling and delivery regimes that satisfy consumer demands regarding food quality, safety, health and the environment.

This outlook provides a concise summary of the critical factors that contribute to the successful and equitable integration of developing countries' producers into sustainable (inter)national agro-food chains and networks. More precisely, we address in the remainder of this article the following issues:

• the conditions for successful integration of producers from the South into (inter)national agro-food chains and networks

- the role and contributions of agro-food chains and networks for developing market opportunities for smallholders in the South
- the institutional, governance and contractual requirements for meeting the growing number of grades and standards
- the roles and contributions of public–private partnerships and knowledge institutions to support sustainable agro-food chains and networks.

Following the more conceptual discussion presented at the introduction of this book, we will now refer to the contributions from the field and the business cases presented at the conference for practical illustration of the arguments.

Do cross-border food chains and networks matter?

Supply-chain integration can be an important vehicle for providing access to remote markets, enabling producers and processors to respond to (changes in) consumer demand and facilitating joint innovation and upgrading. The business cases presented at the conference provide evidence that such partnerships are indeed effective, in line with experiences documented elsewhere (see for example the Proceedings of the International Conferences on Chain and Network Management in Agribusiness and the Food Industry organized from 1994 to 2004 by Wageningen University; also Vellema and Boselie 2003; Van der Vorst 2000; Claro 2004; Camps et al. 2004). Most important functions of supply-chain integration are:

- food chains and networks can be helpful to reduce transaction costs (see the business cases on medicinal plants in India beef in Brazil and fruit from South Africa)
- food chains contribute to enhance quality (as documented in the business cases of fruit exports from South Africa and local vegetables sourcing for TOPS supermarkets in Thailand)
- food chains enhance the sustainability (see the experiences of banana exports from Peru, cocoa in Costa Rica and medicinal plants in India)
- food chains and networks could reduce uncertainties regarding market outlets (see medicinal plants in India and *Allanblackia* in Ghana)
- participation in the chain supply may create wealth (see the business cases of beef in Brazil and Fresh Partners in Thailand).

Supply-chain cooperation thus offers potentially many advantages compared to buying and selling at the open market. However, these advantages cannot be reaped without major costs and efforts (see business cases of fish in Kenya and vegetable sourcing by TOPS in Thailand). The main lesson derived from the business cases is that the high variability in size and quality of the produce poses serious limitations for integrating long-distance or cross-border food chains that are able to achieve development objectives such as the inclusion of smallholders (e.g. compare the performance of TOPS and Fresh Partners in Thailand). In addition, smallholder participation is particularly favoured by trade arrangements that guarantee permanent market access (like in the fair-trade banana exports from Peru and the cacao exports from Costa Rica). The international business environment tends to be highly competitive and therefore economic returns are sometimes rather poor and not evenly distributed over all the actors in the food chain. Therefore, investments in supply-chain upgrading can only be expected when real partnerships are established (like the Freshmark sourcing in South Africa or the preferred-supplier arrangements of Hortfruta in Central America; see contributions by Van Deventer and Reardon in this volume). The main question derived from these experiences is, therefore, under which conditions successful integration of agro-food chains and networks is likely to take place and how supply chains can function as an instrument for development. We address this question by discussing both global and local issues that influence the emergence of inclusive agro-food chain development.

Global issues

From a theoretical point of view, several studies have revealed the mechanisms that could enhance partnerships in food chains and networks. The fundamental conditions for vertical cooperation in chains as forwarded by Williamson (1989) emphasize the need for reducing transaction costs. Companies may refrain from involving in open market transactions, even where actors are independent, have free choice to exchange and do no have authority over each other. This is explained by the fact that market exchange mechanisms are not costless, due to bounded rationality of agents, the occurrence of asymmetric information or the requirements to invest in specific assets that 'lock in' farmers or the company in specific relationships.

Consequently, dairy farmers are operating at the mercy of the processing factory once their cows start producing milk, and they will try to reduce their uncertainty by arranging delivery contracts for selling the milk. Similarly, agro-food companies may own plantations in order to assure the supply of commodities. The costs of these internal transactions may be relatively high compared to the open market where the company could buy from whomever and whenever. However, the certainty of supply and the possibilities to enforce specific product standards could lead them to prefer contractual exchange.

In theory, there are three types of governance structures: (1) open market delivery, (2) contracts (see the Shoprite case in South Africa in the contribution by Van Deventer; also the TOPS sourcing system in Thailand) and (3) hierarchy based on vertical chain control (as illustrated in the case of Brascan beef in Brasil). Pure market 'chains' consist of independent partners that decide at every occasion whether they will engage in the exchange of goods and services. This is only feasible for undifferentiated commodities. Contracts are agreements in which the buyer (trader or processor) co-invests in the production, for instance by providing seeds and credit and by describing desired agricultural practices. The buyer guarantees the purchase, which decreases market uncertainty for the producer.

Finally, in hierarchical types of supply-chain governance, the buyer secures supply by directly owning and operating production facilities. Chain control and steering at all levels is thus executed by the buyer, who incurs relatively high internal transaction costs.

In practice we observe an increasing number of agro-food supply chains where reciprocal relationships have been established. In many cases, informal norms are guiding the behaviour of buyers and sellers and even precede formal contracts. It is also shown that the bundle of governance mechanism may differ depending on the development stage of the agro-food supply chain (see Table 1). Food chains that are in an early stage of development rely on basic information exchange and look for attuning some logistic processes or establishing quality codes (like in the Nile-perch business case in Kenya). In some more advanced settings, actors start investing in joint marketing efforts, engage in some research and development, and invest in fixed assets such as processing facilities (e.g. Brascan beef in Brasil; vegetable stations in Thailand). In such circumstances, switching costs tend to increase since agents in the chain become more interdependent.

A second global issue refers to the chain environment, particularly the distribution of costs and benefits between public and private agents. This is especially important when externalities are involved. Fair trade and ecological labels (see business cases of banana in Peru and cacao in Costa Rica) aim to incorporate externalities in the price. New food chains (fish from Kenya, medicines from India, *Allanblackia* from Ghana) are able to generate income in the short term but also face the risk of unsustainable exploitation of the environment, which represents an implicit public cost. Most private partners involved in these chains can either continue until depletion and then move to another place or share this concern regarding the externalities when they want to guarantee a long-term resource base for their raw materials (see keynote presentation by Bordewijk). In the cases of medicine and *Allanblackia* production, the private partners assumed their co-responsibility by taking proactive measures to broaden and sustain the resource base through investments in technologies for cultivation of formerly wild products. Cooperation with knowledge institutions (i.e. KIT, SNV, Wageningen UR) in such partnerships proved to be of key importance for enhancing technology development and to guarantee brokerage between local knowledge and industry demands. In a similar vein, the risks of (over)fishing in Lake Victoria could be reduced by measures aiming at reducing the large waste in the food chain. Long-term investments of this kind can only be expected if reliable and sustainable partnerships are established and specific governance structures (public grades and standards; see contribution by Reardon) are put in place.

Conditions for sustainable supply-chain integration

Institutional and macroeconomic factors have been mentioned as being of critical importance for establishing food chains and networks that are capable of engaging in cross-border exchange. Rodriguez in his contribution referred particularly to the importance of cutting down price-distorting subsidies, while Van der Meer and Reardon emphasize the role of non-tariff barriers related to sanitary rules and quality standards. Import barriers, tariffs and non-tariff policies may impede the access to export markets, as illustrated most clearly by the EU import regimes for banana (see presentation by La Cruz). When import restrictions of whatever nature exist, then chances for entering new markets decrease rapidly. Local policies can also lead to market distortions; Oyewole provides an example of discriminatory credit supply that inhibits the development of the cassava-processing industry in Nigeria.

Market organization also plays a major role in the establishment of integrated supply chains. If the market structure tend towards oligopoly conditions – typical for the retail industry in some countries – the access to these market outlets becomes difficult. Also, if competition in a particular market is cutting-edge, the chances for successful entry of newcomers are relatively limited, unless some particular attributes are offered, such as lower prices or unique qualities (e g. fair-trade banana from Peru and ecological cacao from Costa Rica).

An additional factor influencing market access refers to the governance regimes maintained on both sides of the supply chain. Good governance may appear as a non-tariff barrier when the buyer puts forward specifications regarding child labour, good labour relationships, sustainable production practices in the form of good agricultural practices (GAP), respect for human rights, just to mention a few. If sellers cannot comply with these specifications, or do not fully comply with these demands, they may jeopardize their position in the food chain. Institutional codes based on public policies, such as the *Codex Alimentarius* (FAO) or derived from private arrangements (like ISO, Eurepgap, BCR and many others) may impede newcomers to deliver successfully to foreign market outlets (see contributions by Fresco and Reardon). Therefore, to participate in these more demanding markets, investments in product and process upgrading are a prerequisite. These barriers may partly be overcome if the chain partners invest in training programmes of smallholder farmers that enable them to comply with the standards in order to remain included in the supply chain (see examples of quality training provided by Fresh Partners in Thailand and GAP trainings offered to farmers by Unilever).

Beyond the macroeconomic and institutional factors, a wide range of microeconomic and management factors can be identified that influence the prospects for chain and network cooperation. Such cooperation will only arise if the expected and achieved economic and social return for engaging in supply chains and networks are larger than the costs. Farmers are not likely to become involved in contractual deliveries if the costs and/or the risks exceed the potential benefits. Similarly, food-processing companies will not engage into upstream relationships with farmers if the costs of doing so are not in balance with expected returns in terms of volumes, quality and price. The trade-offs between costs and benefits are directly related to the capabilities of actors within the food chain or network, since the weakest link may jeopardize the investments of others. Therefore, supply-chain development is inherently related to capacity building in several directions: technical skills, economic capacities and social and managerial experience related to production, processing, marketing, logistics etc. Without these capabilities in place, the agro-food chain as a whole will easily suffer from disintegration. Building capacity and capabilities within the supply chain are investments to create successful agro-food networks. Several examples of successful capacity building are provided in the business cases included in this volume.

A prerequisite for joint investment by supply-chain agents is the existence of certain coherence in values and objectives amongst the stakeholders. If the actors involved in the food chain are not focused on the same objectives, investments are in vain. A key factor that has been identified in many cases was the creation of trust. Trust can be considered the cornerstone for building relationships; the establishment of integrated food chains depends essentially on building trust and reciprocity (Ostrom and Walker 2003; Migchels 2001). As explained by Lewis (1999), building up trust depends on the conditions for establishing trust, the practices that earn trust and the safeguards encouraging trust. In many cross-border food chains, the proximity factor is crucial as a condition for trust: proximity refers not only to physical factors, but particularly to cross- cultural communication. However, trust and control are two related processes: the more efforts chain actors put on control mechanisms, the higher the chances for shirking behaviour. However, absence of control mechanisms may invite to free-rider behaviour. Balancing both aspects asks for particular management arrangements and organizational regimes. The business cases provide conclusive evidence of the positive contribution of involving third parties for facilitating the design, establishment and maintenance of agreed business practices in the food chain. Third parties may be helpful to overcome or modify imbalances in bargaining power and could create the necessary conditions for enhancing trust. In addition, due attention needs to be given to entrepreneurship as the driving force for developing food chains and networks: supply-chain integration needs to be based on business strategies that nurture entrepreneurship: it is not for free!

How to organize successful integrated supply chain?

The organization of stakeholders into (inter)national agro-food chains and networks is likely to follow particular pathways. Building effective partnerships requires initially strong efforts for streamlining production processes and handling practices, while at later stages efforts could be made towards chain upgrading or new product development. Based on a cross-section comparison of the business cases presented at the conference, we can identify a number of common issues and organizational challenges that are typical for different stages of supply-chain and network integration (see Table 1).

Even while in practice the development of supply chains involves a multitude of dynamic issues that may coincide in time, we may consider a simple description of the life cycle of typical food chains and networks. In each stage, particular demands and challenges regarding internal relationships and external positioning become apparent (see Table 1). Examples of supply chains in an initial phase are provided in the business cases of medicines in India and *Allanblackia* in Ghana. The Nile-perch business from Lake Victoria and the cocoa cooperatives in Central America are currently in the organization phase. Fruit from South Africa and the Freshmark case (in the presentation by Van Deventer) are good examples of business at the implementation stage. Finally, Brascan beef in Brasil and TOPS vegetables sourcing in Thailand are in the optimizing stage; the latter firm has recently been sold to another agent[1].

Table 1. *Typical challenges at different stages of supply-chain cooperation*

Stage	Critical issues	Organizational challenges
Initial initiative	Looking beyond business to market transactions. Strategic objective setting. Partner assessment and selection.	Trust building, building informal rules for behaviour (with involvement of third party for strengthening organization).
Organization stage	Defining the competitive position of the food chain and the associated competitive strategy. Distribution of margins and implementation of control functions. Allocating of risks.	Establishment of internal governance structures (organizing procedures, division of tasks and responsibilities) Building of trust. Involvement of third party for business management training.
Implementation stage	Focus on 'making it work'. Monitoring of activities and results. Procedures for conflict resolution.	Maintaining trust; building loyalty. Increase of information exchange and knowledge sharing. Specification of procedures.
Optimizing stage	Improving quality of products, the required processes for upgrading and the organization of the partnership.	Maintain and reinforce trust. Research and development (R&D). Labelling and branding.
Decline stage	Exit strategies (merger or take-over).	Step-by-step or abrupt dissolution of partnerships.

One of the common issues at all stages refers to the distribution of information, risks and returns. Although so-called open-book calculations may be shared amongst the stakeholders in the chain, in many cases information sharing is postponed to later stages in the life cycle when more confidence is available. In some settings, third parties may be involved to assist this process, assuming a role as initiator or facilitator of supply-chain cooperation. Their activities are initially concerned with fostering cooperation, but may be devoted to capacity development, training and even product development in subsequent stages. The business cases of medicinal plants in India and *Allanblackia* in Ghana provide examples of the involvement of intermediate partners that assist to design the chain and address some of the problems related to trust, support farmers' organization and assist in the development of new technologies.

During the initial and organization stages, the main attention is devoted to activities that permit chain agents to strengthen the internal procedures and practices and to reinforce governance regimes. In addition, partners will engage in a process of strategic assessment to identify their strong and weak points, thus enabling them to define their potential market position. At the subsequent implementing stage, many new and unforeseen problems may be met that put pressure on the participants and lead to a continuous demand for information (see Nile-perch business case of Kenya[2]). Due to information asymmetries, conflicts may arise easily and induce a need for conflict resolution procedures and control mechanisms (as illustrated in the Fruitful case). When entering the optimizing stage, the responsiveness of the food chain to demands from buyers and vis-à-vis competitors is at stake. Improvements in logistical systems become important, while in some other cases product improvements are necessary. An example is found in the Brascan beef case from Brazil, where stakeholders aim at higher profits through horizontal expansion (larger volumes) and through vertical integration by incorporating chain partners such as slaughterhouses and feed-producing companies into the business. Another example are the improved vegetable-sourcing regimes in Thailand that aim at higher quality and food safety criteria for supermarket procurement and export market deliveries.

Once partners find that the value creation in a particular food chain decreases relative to other modes of entrepreneurial opportunities, these partners may abruptly or step-by-step work towards dissolving the partnership. In cases where switching costs are high, barriers to exit can be prohibitive and exit strategies will be deployed even by force, law or through acquisition by partners in the food chain. Typical examples of supply-chain breakdown are given in the business case on cacao from Costa Rica, where the processing company went bankrupt while leaving the other partners in the chain without a market. In a similar vein, the market deregulation of the fruit in South Africa allowed market access of some lower-quality producers that rapidly spoiled the market (trust) that existed before. The international fruit network, consisting of several interlinked supply chains, had to be fully restructured. In this case, the breakdown was followed by a stage of redesign of the supply chain.

Mechanisms for sustainable food chains and network integration

Cooperation of producers, processors, traders and retailers within a setting of supply-chain integration is by no means an easy task. It is therefore highly important to identify some simple mechanisms that proved to be helpful in practice to enhance sustainable agro-food chain and network integration:

- Reduce complexity: supply chains that involve a large number of very heterogeneous participants are likely to face many coordination problems. Involving a larger number of smallholder producers puts high demands on the facilities for sharing information, for reaching agreements of mutual consent, for monitoring processes and for managing the chain. Complexity also increases in cases where some partners have to make larger investments than others; such investment asymmetry puts pressure on the distribution of rents and risks, adding up to the already existing complexities in decision making. Another source of complexity refers to multiple objectives (particularly in emerging new food chains) that can lead to diverting efforts and energy. To avoid this situation calls for restrictive behaviour and controlled ambitions by each of the chain partners.

- Starting at home. It may be challenging to start up a cross-border food chain, but it tends to be better to start operations in nearby markets. If one is not able to develop a food chain for the domestic market, it is highly unlikely that engagement in cross-border chains will be successful. This may be true in general; however, some particular conditions may prevail that enable cross-border food-chain development. Typical examples are the East-African flower industry and labelled food products for particular market segments. Most certified products need to be exported as they aim at consumers with a high purchasing power that are willing to pay additionally for environmental or social aspects (e.g. ecologically produced cacao can hardly be sold in Costa Rica ; neither is there a large market for fair-trade bananas in Peru).

- Farmers' organization. An important aspect for reaching scale concerns the way of organizing primary producers, farmers or growers. Some proponents of rural cooperatives have achieved good results (as illustrated by the cases of cocoa farmers in Costa Rica and banana farmers in Peru), but in some other occasions, farmers show strong resistance against cooperation (see case study of Kenya) basically due to limited real participation and prevailing risks of corruption. Supply-chain integration increasingly relies on preferred-supplier arrangements (see contribution by Reardon) with a selected number of farmers that are recognized as major suppliers. There is a variety in organizational arrangements that may inhibit or encourage the position of farmers in food chains. Some kind of coordination is required to facilitate effective training, provision of inputs and quality control.

- Incentive structure. A major point of debate is always the distribution of incentives in food chains. This refers to the question how much each agent receives from the total value-added. In practice, most discussions are centred around prices and margins, but this is a rather narrow perception of rewards. The incentive structures deal with price, bonuses, cost-sharing, risk mitigation, short-term and long-term benefits that are part of the value captured by partnerships. Food chain development should consider all the different components of the incentive structure. If price motives dominate the behaviour of individual actors, short-term objectives tend to prevail and cooperation may easily break down[3].

- Information transparency. Since we cannot expect actors in the food chain to inform all other stakeholders about their operations, information asymmetries are inherent part of the food chain. However, this does not provide an excuse for hiding critical information in the chain. Building up trust amongst supply-chain participants requires sharing and disclosing information. For some processes, such as tracking and tracing systems, information transparency is a prerequisite. In the cocoa chain an insurance fund has been created to compensate farmers for product denial (contaminated cacao with pesticide residues), asking full openness regarding the causes of such underperformance. Similarly, it is necessary to protect farmers who deliver certified (fair-trade or organic) products against free-rider behaviour of conventional farmers and against individuals that want to hide their behaviour for others.

- Exchanging experiences. A last mechanism that may induce food chain partnerships, is based on sharing experiences from others agents, like organizing assistance from supporting agencies, market brokers or knowledgeable institutes. Worldwide there is a wealth of experience regarding supply-chain integration and cooperation that has to be documented and can be provided to interested partners engaged in supply-chain programmes for development.

Challenges for research and policy

The papers and business cases presented at the Wageningen conference and workshop on agro-food chains and networks for development permit to draw a number of more general conclusions regarding the challenges for research, the private sector and policy makers.

Researchers showed that their research efforts can contribute to a better understanding of the structure and dynamics of agro-food chains and their potential contribution to enhancing development and poverty alleviation. In many of the presented business cases, it became clear that researchers do not just analyse food chains but also can play an intermediate and facilitating role for chain upgrading through their close interactions with all food chain partners. Research can therefore contribute to addressing the challenge of supporting the development of effective agro-food chains. Another major task for research institutes is to assist in the design of improved supply-chain processes, by fostering technologies that fit customers' demands and serve the interests of chain participants. Finally, at more aggregate macroeconomic and policy level, research can contribute to identify policy devices that reduce trade restrictions and favour sustainable market access to producers from the South..

For professionals in the public sector, development of agro-food chains and networks offers a new challenge to policy making. In the past, most attention has been given to strategies for finding market outlets without direct involvement of the private sector. Nowadays, national policies are increasingly designed while considering incentives for the development of food chains. Due attention needs to be given to improve the effective and impartial operation of public-service agencies (customs, inspection agencies, port authorities etc.) to reinforce the network in which food supply chains are embedded. Another major challenge for policy makers is how to tailor generic policies to the specific demands forwarded by internal and international food chains and network agents. Development policies aiming at enhancing trade to provide livelihood to rural poor should be brought in line with public-health policies imposing barriers to trade for products considering food safety requirements.

For professionals from the business sector, the challenge is to share their expertise in building up food chains, learning how these processes work in the particular context of developing countries, and assess the promising approaches that give the best result. Professional staff facilitating the development of food chains may benefit much from the exchange of experience with representatives of farmers' organizations and NGOs working in the South, and each of them might derive clear benefits from participation in such platforms.

In summary, the exchange of ideas between experts from food chain partners, including representatives from producer organizations in the South, non-governmental organizations, policy makers, private-sector parties and research institutions proved to be highly valuable for increasing our understanding of the common and differentiated interests in agro-food supply-chain integration. The participants created value by sharing their experiences in the debate, and, in the discussions on various business cases, they were confronted with advice and comment from the audience that challenged them to reconsider or further improve their approach. Meetings between practioners and professionals from such different backgrounds can serve as a market place that facilitates new partnerships and will hopefully lead to new initiatives.

NOTES

[1] The Thai Fresh project passed essentially all four stages and explicitly acknowledged legal access to markets as a key factor in the initial formation stage, institutional access to markets in the organization stage, trust as key issue during the implementing stage, and risk sharing and return during the optimizing stage.

[2] In the case of the fish supply chain from Lake Victoria, Kenya, all aspects of the chain work and profit is made, but a lot of fish is lost during handling operations, occasioning low efficiencies and an unnecessary high pressure on the remaining fisheries stock. Optimization of handling to reduce losses can decrease the number of fish to be caught and support a more sustainable use of the fish reserves, and at the same time increase profit of the fish factory as most of the fish entering can be fully transformed to valuable end products.

[3] In the fair-trade banana chain from Peru and the ecological-cacao chain from Costa Rica, the certification and related premiums are distributed through cooperatives and part of the revenues are kept at collective level, for instance as an insurance fund to cover costs related to product denial, unexpected natural disasters or to provide scholarships to the associated farmers.

REFERENCES

Camps, T., Diederen, P., Hofstede, G.J., et al., 2004. *The emerging world of chains and networks: bridging theory and practice*. Reed Business Information, 's-Gravenhage.

Claro, D.P., 2004. *Managing business networks and buyer-supplier relationships: how information obtained from the business network affects trust, transaction specific investments, collaboration and performance in the Dutch potted plant and flower industry*. Proefschrift Wageningen [http://library.wur.nl/wda/dissertations/dis3527.pdf]

Lewis, J.D., 1999. *Trusted partners: how companies build mutual trust and win together*. The Free Press, New York.

Migchels, N.G., 2001. *The ties that bind: a dynamic model of chain cooperation development*. Proefschrift Technische Universiteit Eindhoven [http://lx1.library.wur.nl/wasp/bestanden/LUWPUBRD_00121964_A502_001.pdf]

Ostrom, E. and Walker, J., 2003. *Trust and reciprocity: interdisciplinary lessons from experimental research*. Rusell Sage Foundation, New York.

Reardon, T. and Timmer, C.P., in press. Transformation of markets for agricultural output in developing countries since 1950: how has thinking changed? *In:* Evenson, R.E., Pingali, P. and Schultz, T.P. eds. *Handbook of agricultural economics. Vol. 3: Agricultural development: farmers, farm production and farm markets*. Elsevier, Amsterdam.

Van der Vorst, J.G.A.J., 2000. *Effective food supply chains: generating, modelling and evaluating supply chain scenarios*. Proefschrift Wageningen [http://www.library.wur.nl/wda/dissertations/dis2841.pdf]

Vellema, S. and Boselie, D., 2003. *Cooperation and competence in global food chains: perspectives on food quality and safety*. Shaker Publishing, Maastricht.

Williamson, O.E., 1989. Transaction cost economics. *In:* Schmalensee, R. and Willig, R.D. eds. *Handbook of industrial organization*. North-Holland, Amsterdam, 3-59.

WAGENINGEN DECLARATION

TOWARDS SUSTAINABLE INTERNATIONAL AGRO-FOOD CHAINS AND NETWORKS FOR DEVELOPMENT

At the occasion of the opening of the academic year at Wageningen University on 6 September 2004, representatives from farmers' organisations, public agencies and private enterprises from the North and the South gathered to discuss the potentials and challenges of international agro-food chains and networks for enhancing sustainable development.

Further liberalisation of international trade does not guarantee access to markets by smallholders from the South. Quality demands and phytosanitary regulations tend to impose new restrictions. Therefore, joint actions are required to enhance incentives for investment and innovation, to reinforce the business climate in developing countries, to improve competitiveness of local farms and firms, and to nurture the development of local entrepreneurship.

Efficient and equitable supply chains for agro-food products can substantially benefit from stable networks and mutual relations between producers, processors, traders and retailers. Public-private partnerships are vital for capacity development and shaping adequate access conditions. Knowledge institutions and voluntary organisations can facilitate co-innovation between chain partners and support synergies between public and private partners.

The following critical issues and challenges have been identified:

- To develop market outlets for products from the South
- To enhance business-to-business relationships between suppliers from the South and customers from the South and the North
- To enhance capacity with local producers and increase economies of scale
- To improve access to market information and logistical systems
- To support the upgrading of quality and safety of agro-food products with particular attention to smallholders
- To develop new products with higher value-added in developing countries
- To identify prospects for branding, labelling and certification of products
- To build transparency and trust within intercultural settings.

The participants of the Wageningen Conference on Agro-Food Chains and Networks for Development call upon governments and the entrepreneurial sector to join forces in a partnership programme for co-innovation towards sustainable food chains in order to contribute to reducing poverty and unemployment in developing countries and bringing the world closer to the Millennium Development Goals.

Wageningen UR Frontis Series

1. A.G.J. Velthuis, L.J. Unnevehr, H. Hogeveen and R.B.M. Huirne (eds.): *New Approaches to Food-Safety Economics.* 2003
ISBN 1-4020-1425-2; Pb: 1-4020-1426-0
2. W. Takken and T.W. Scott (eds.): *Ecological Aspects for Application of Genetically Modified Mosquitoes.* 2003
ISBN 1-4020-1584-4; Pb: 1-4020-1585-2
3. M.A.J.S. van Boekel, A. Stein and A.H.C. van Bruggen (eds.): *Proceedings of the Frontis workshop on Bayesian Statistics and quality modelling.* 2003
ISBN 1-4020-1916-5
4. R.H.G. Jongman (ed.): *The New Dimensions of the European Landscape.* 2004
ISBN 1-4020-2909-8; Pb: 1-4020-2910-1
5. M.J.J.A.A. Korthals and R.J.Bogers (eds.): *Ethics for Life Scientists.* 2004
ISBN 1-4020-3178-5; Pb: 1-4020-3179-3
6. R.A. Feddes, G.H.de Rooij and J.C. van Dam (eds.): *Unsaturated-zone Modeling.* Progress, challenges and applications. 2004   ISBN 1-4020-2919-5
7. J.H.H. Wesseler (ed.): *Environmental Costs and Benefits of Transgenic Crops.* 2005   ISBN 1-4020-3247-1; Pb: 1-4020-3248-X
8. R.S. Schrijver and G. Koch (eds.): *Avian Influenza.* Prevention and Control. 2005   ISBN 1-4020-3439-3; Pb: 1-4020-3440-7
9. W. Takken, P. Martens and R.J. Bogers (eds.): *Environmental Change and Malaria Risk.* Global and Local Implications. 2005
ISBN 1-4020-3927-1; Pb: 1-4020-3928-X
10. L.J.W.J. Gilissen, H.J. Wichers, H.F.J. Savelkoul and R.J. Bogers, (eds.): *Allergy Matters.* New Approaches to Allergy Prevention and Management. 2006
ISBN 1-4020-3895-X; Pb: 1-4020-3896-8
11. B.G.J. Knols and C. Louis (eds.): *Bridging Laboratory and Field Research for Genetic Control of Disease Vectors.* 2006
ISBN 1-4020-3800-3; Pb: 1-4020-3799-6
12. B. Tress, G. Tress, G. Fry and P. Opdam (eds.): *From Landscape Research to Landscape Planning.* Aspects of Integration, Education and Application. 2006
ISBN 1-4020-3979-4; Pb: 1-4020-3978-6
13. J. Hassink and M. van Dijk (eds.): *Farming for Health.* Green-Care Farming Across Europe and the United States of America. 2006
ISBN 1-4020-4540-9; Pb: 1-4020-4541-7
14. R. Ruben, M. Slingerland and H. Nijhoff (eds.): *The Agro-Food Chains and Networks for Development.* 2006   ISBN 1-4020-4592-1; Pb: 1-4020-4600-6

Lightning Source UK Ltd.
Milton Keynes UK
22 October 2010

161687UK00003B/29/A